Cell Metabolism: Growth and Environment

Volume II

Editor

T. A. V. Subramanian
Professor
Department of Biochemistry
Vallabhbhai Patel Chest Institute
University of Delhi
Delhi, India

CRC Press, Inc.
Boca Raton, Florida

Library of Congress Cataloging-in-Publication Data
Main entry under title:

Cell metabolism: growth and environment.

Bibliography: p.
Includes indexes.
1. Microbial metabolism. 2. Microbial metabolites.
3. Microbial synthesis. 4. Plants—Metabolism.
5. Metabolism—Regulation. I. Subramanian, T. A. V.
QR88.C45 1986 574.87′5 85-28966
ISBN 0-8493-5750-0 (v. 1)
ISBN 0-8493-5751-9 (v. 2)

Direct all inquiries to CRC Press, Inc., 2000 Corporate Blvd., N.W., Boca Raton, Florida, 33431.

International Standard Book Number 0-8493-5750-0 (v. 1)
International Standard Book Number 0-8493-5751-9 (v. 2)

Library of Congress Card Number 85-28966
Printed in the United States

PREFACE

The economic importance of several microbial metabolites, particularly antibiotics, has provided impetus to research efforts to elucidate the genesis of these compounds. The characterization and identification of biosynthetic routes has greatly facilitated the creation of newer and more efficient microbial products. The main difficulty is the tight degree of metabolic regulation which is frequently encountered in microorganisms, rendering them inefficient producers of metabolites. A certain degree of relaxation of the metabolic regulatory mechanisms is essential for economical production of an abundance of the desired microbial product. In other words, for the microbial cells to serve as efficient producers it is very important to circumvent or overcome the regulatory mechanisms involved. In the competitive world of commercial synthesis and production, simple molecules like glutamic acid and lysine are still made by fermentation rather than by chemical synthesis. About 300,000 tons of citric acid are produced annually from *Aspergillus niger* alone. In addition, microorganisms are extremely useful for carrying out bioconversion processes involved in the production of specific stereo compounds.

Increased biosynthesis of chlortetracyclin by developing mutants of *S. aureofaciens* resistant to the antibiotic has recently been reported. With the developments in biotechnology it is possible now to manufacture biologically important compounds like insulin on a large scale without depending on animal sources. Some of the hydroxylation reactions in steroid biosynthesis which require multistep procedures in chemical synthesis are greatly facilitated by microbial enzymes.

Historically, microbial products have been divided into primary and secondary metabolites. Primary products like proteins, nucleic acids, and carbohydrates are essential for growth and metabolism. Secondary products until recently were believed to have no significant role in the physiology of the producing organism and were considered as metabolic waste products.

As a first step it is necessary to find a wasteful strain which will overproduce and excrete a particular compound. The next stage is to modify the regulatory control of the strain to further increase its inefficiency. An example of such exploitation of microorganisms to waste their precious metabolites is seen in riboflavin production (up to 5000 $m\ell/\ell$) by commercial cultures of *E. ashby*.

These two volumes deal with the regulation of metabolism in microbes and plants with respect to differentiation, proteins, nucleic acids, aromatic acids, alkaloids, and other secondary metabolites like aflatoxins and streptomycin.

T.A.V. Subramanian
Vallabhbhai Patel Chest Institute
University of Delhi
Delhi-110007, India

THE EDITOR

Dr. T. A. V. Subramanian holds a Ph.D. in Biochemistry from the University of Madras. He has been Head of the Department of Biochemistry, V. P. Chest Institute, University of Delhi, since 1956.

Dr. Subramanian's training and experience, which includes 2 years at Columbia University, New York and 2 years at University of Wisconsin, Madison, is working in areas of biochemistry connected with metabolism of *Aspergillus parasiticus*, metabolism of various strains of tubercle bacilli, and pulmonary surfactants. He has more than 250 publications in these areas. He has been actively engaged in research on various aspects of the metabolism of *Aspergillus parasiticus* for more than 20 years. He has been President of the Society of Biological Chemists (India).

CONTRIBUTORS

Zia Uddin Ahmed
International Centre for Diarrhoeal Diseases Research, Bangladesh (ICDDR,B)
G.P.O. Box 128, Dhaka-2
Bangladesh

Kenji Ishii
Department of Toxicology and Microbial Chemistry
Science University of Tokyo
Tokyo, Japan

Elmer H. Marth
Professor of Food Science and Bacteriology
University of Wisconsin-Madison
Madison, Wisconsin

Zdeněk Řeháček
Head of the Laboratory of Microbial Metabolites and Cell Differentiation
Institute of Microbiology
Acting Director of the Czechoslavakian Academy of Sciences International Postgraduate Courses UNESCO
Prague, Czechoslovakia

Yoshio Ueno
Professor
Department of Toxicology and Microbial Chemistry
Tokyo University of Science
Tokyo, Japan

Leo C. Vining
Professor
Department of Biology
Dalhousie University
Halifax, Nova Scotia
Canada

Ahmed E. Yousef
Research Associate
Department of Food Science
University of Wisconsin-Madison
Madison, Wisconsin

CELL METABOLISM: GROWTH AND ENVIRONMENT

Volume I

Volume II

TABLE OF CONTENTS

Chapter 1

BIOSYNTHESIS OF POLYKETIDES

Ahmed E. Yousef and Elmer H. Marth

TABLE OF CONTENTS

I. INTRODUCTION

Polyketides are a group of biogenically related natural compounds that are usually produced by fungi. Some of the early discoveries about polyketides date back to the turn of the century, but our knowledge about this class of compounds is still far from complete. Polyketides include such highly hazardous compounds as aflatoxins, which have shaped and will continue to shape the agricultural economy in the world for years to come.

Workers in this field have developed a continuously growing list of hazardous metabolites produced by molds, but excluded are organisms from the ever-diminishing list of so-called useful fungi. Hence, researchers working in this field share an important responsibility, and the need exists for intensified work that will lead to a better understanding of fungi and their secondary metabolism.

II. SPECTRUM OF POLYKETIDES

In the broadest sense, polyketides may be defined as those diverse natural products that are related by their common biosynthetic pathway, viz. the acetate-polymalonate route.[1,2] Acetate or other acyl units, such as propionate, benzoate, or cinnamate, can act as chain-initiating moieties, and beside malonate, though less frequently, propionate or butyrate can act as chain-elongating units.[2] This broad definition of polyketides suggests that a wide variety of products such as fatty acids, polyacetylenes, phenols, macrolides, flavonoids, and others are covered by the term. Because of the broad spectrum of compounds included in this definition, many workers in this field prefer to restrict the term to those compounds that are biosynthesized via the acetate-polymalonate pathway, but without obligate reduction of intermediate compounds.[3-5] This restriction sets fatty acids apart as a distinctive group of compounds. This chapter will use the restricted definition of polyketides, with occasional reference to biosynthesis of fatty acids when comparison is necessary. Even in this narrow sense of the term, polyketides are a very diverse group of compounds that vary in complexity from simple tetraketides, such as orsellinic acid, to compounds of elaborate structure, such as the decaketide aflatoxin B_1 (Figure 1).

III. OUTLINE OF BIOSYNTHESIS OF POLYKETIDES

Biosynthesis of polyketides[1,3,4] involves condensation of an enzyme-bound acyl (mostly acetyl) moiety as the chain-initiating unit with three or more of malonyl (chain-elongating unit). During the condensation, malonyl groups are decarboxylated, adding "C_2-units" to the enzyme-bound ketoacyl intermediate compound. Depending on the number of C_2-units that constitute the fully assembled β-polyketoacyl intermediate compound, polyketides are divided into tetraketides, pentaketides, hexaketides, etc. For the simplest tetraketide, orsellinic acid, the enzyme-bound intermediate compound cyclizes and leaves the enzyme surface without further modification (Figure 2). However, this is not true for most polyketides, which usually undergo various degrees of secondary modifications including alkylation, oxidation, reduction, ring cleavage, etc. These modifications may occur on the enzyme-bound intermediate compound or after the intermediate compound leaves the enzyme surface. It should be emphasized at this point that the scheme just presented is speculative, and is based on a recent version of the so-called "polyketide hypothesis".[6] A full account of the scheme for biosynthesis of polyketides will appear later in this chapter.

IV. METHODOLOGY TO STUDY THE BIOSYNTHESIS OF POLYKETIDES

The study of polyketides, although including peculiar features, seeks answers to the same basic questions asked before any metabolic study is done. The questions to be answered by

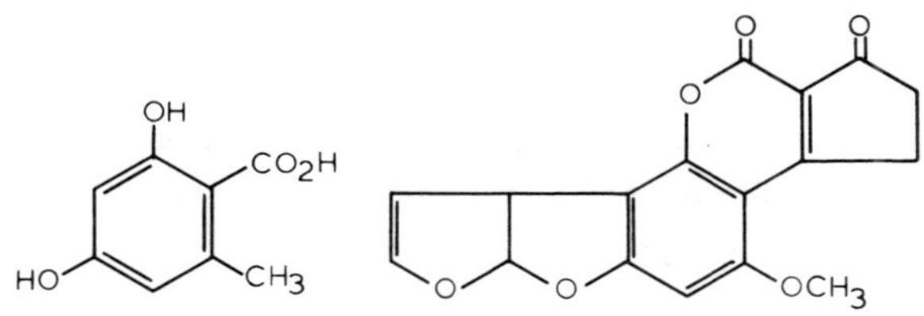

FIGURE 1. Structure of (A) orsellinic acid and (B) aflatoxin B_1.

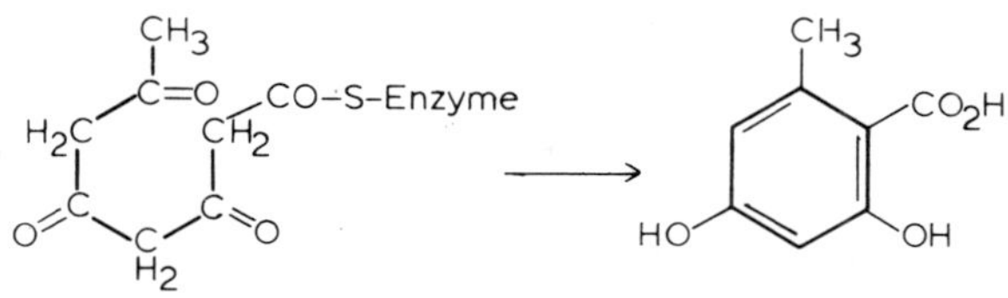

FIGURE 2. Cyclization of orsellinic acid intermediate compound.

such studies are: What are the building blocks? What are the intermediate steps? By which mechanism(s) are these changes brought about? To pursue such a study there should be available: (1) a rich source of biosynthetic activity, (2) the putative precursor, intermediate compound, or an even more remote substrate, and (3) a method to detect and characterize the end product and sometimes the precursor or intermediate compound(s). We will discuss these requirements in more detail as related to studies of polyketides.

A. Biosynthesizing System

To study a biosynthetic route, one first needs a rich source for the biosynthetic activity that leads to the product under investigation. Since most known polyketides are produced by fungi,[5] examples dealing with polyketide-producing organisms will be limited to these organisms. Early studies usually dealt with the entire mold. As knowledge accumulated and hypotheses were made of the kind of biosynthetic activity involved, attempts to isolate the pure enzymes that catalyze various steps of the pathway followed. Enzymatic evidence for the polyketide hypothesis was reviewed by Light[6] in 1970.

1. Growing Cultures

A pure culture of mold is grown on a suitable medium. Minimal media generally work better than highly nutritious subtrates for studies on secondary biosynthesis. This is true because limitation of a nutrient (which is achieved faster in minimal that in highly nutritious media) is thought to trigger secondary metabolism.[3,7] A medium limited in its nitrogen source is probably the one of choice for the synthesis of polyketides.[8] A growing culture may be used as the ultimate synthesizing system or the culture may be further manipulated by one or more of the purification techniques that will be dealt with later. When used as the ultimate biosynthesis system, a growing culture of mold gives only a crude idea about the pathway, and overinterpretation of the results can lead to great ambiguities. Such a study usually gives a preliminary idea about the environmental factors that regulate the in vivo biosynthesis of the metabolite under investigation.

Use of a growing mold culture to study the biosynthesis of polyketides (and probably the other secondary metabolites) could be more informative if the kinetics of mold growth and of production of the metabolite are clearly demonstrated.[9-11] By this means, segregation between the two overlapping events, growth and secondary biosynthesis, is possible. The growing culture may be fed a labeled precursor or intermediate compound, and the sequence of incorporation of the label into various putative intermediate compounds and the end product is followed. By this technique (called kinetic pulse labeling), the sequence of appearance of intermediate compounds can suggest an overall biosynthetic pathway. Forrester and Gaucher[12] effectively used this procedure during their study on the biosynthesis of patulin.

Usually, liquid media are used to grow molds in studies on the biosynthesis of polyketides. Cultures could be incubated with or without agitation during the course of growth and secondary biosynthesis. The use of fermentors helps to control growth conditions and produces well-phased cultures.

2. Resting Cultures

In a growing mold culture, primary metabolism may severely modify results obtained from the study of secondary metabolism and dilute the label in the final secondary metabolite. To prepare a resting culture, the mold is first grown in a suitable medium until the biosynthetic system is well developed. The mycelium is then transferred to a nitrogen-free medium where growth is arrested. This procedure was used to study the regulation and biosynthetic inhibition of some polyketides.[13,14] The mold suspended in the resting medium could be fully capable of producing the ultimate metabolite or production could be inhibited at a certain step in the biosynthetic pathway through the use of a mutant or a specific enzyme inhibitor. For example, a blocked mutant and dichlorvos-treated *Aspergillus parasiticus* were used to elucidate the biosynthetic pathway of aflatoxin in resting cultures.[15] The intermediate compounds formed immediately before the metabolic block may accumulate in quantities sufficient for isolation and detection. Also, incorporation of an advanced intermediate compound, beyond the blocked step, may be monitored by resumption of the production of the metabolite.

3. Cell-Free Extracts

A further step toward a purer biosynthetic system is to prepare a cell-free extract from a mold culture. The consistent production of polyketides by cell-free extracts was, in many instances, difficult to achieve. This may be related, among other reasons, to the mechanical disruptive technique employed to prepare cell-free extracts.[16] Alternative procedures using no or less severe mechanical disruption were proposed.[16,17] These milder techniques involve enzymatic lysis of the fungal cell walls to set free the fungal protoplast from which the cell-free extract can be easily prepared.

4. Pure Enzymes

Pure enzymes capable of converting different putative intermediate compounds to the metabolite to be investigated constitute unambiguous proof of the biosynthetic pathway. With many polyketides, this goal is extremely difficult to achieve. This difficulty manifests itself by the instability and variability of activity between different enzymatic preparations.[6] However, several enzymes with various degrees of purity have been prepared from polyketide-producing fungi.

B. Precursors or Intermediate Compounds

A carbohydrate such as glucose can be catabolized by mold cells to produce acetate units, which can be viewed as the building blocks of polyketides. Externally supplied acetate also can be incorporated into polyketides. Although the actual biosynthetic route involves malonate as well, the in vivo biosynthesis of polyketides can proceed with acetate being the only precursor supplied. This is true because interconversion of acetate and malonate is a rapid process.[4]

An isotopically labeled precursor of the putative intermediate compound is fed to whole mold cells or added to cell-free extracts to demonstrate specific incorporation of the material into the metabolite being studied. Labeling with ^{14}C was extensively applied to biosynthetic studies of polyketides and labeling with ^{3}H was occasionally employed. Currently, ^{13}C-labeling, accompanied by nuclear magnetic resonance (NMR) spectroscopy, provides a valuable technique for biosynthetic studies of secondary metabolism.

C. End Product Analysis

Metabolic studies can be confined to identifying and/or quantifying the metabolite under investigation or can be expanded to localize certain labeled atom(s) in that metabolite. The degree of sophistication depends on the questions to be answered by the study. Studies on the environmental and regulatory factors of secondary metabolism may require mere quantitation of the metabolite, which does not have to be labeled. Elucidation of biosynthetic pathways, however, usually requires feeding a labeled precursor and determination of the specific incorporation of the label into the product. Detection of the label in the product may be enough to implicate a precursor or intermediate compound in the biosynthetic pathway. Understanding the underlying mechanism, however, requires localization of the label in the product either by chemical degradation (used with ^{14}C-labeling) or NMR spectroscopy (when ^{13}C-labeling is used). About 1 decade ago, double-labeling of precursors with ^{13}C was successfully used as a powerful technique in biosynthetic studies of polyketides.[18] The success of this technique is related to the fact that only those pairs of adjacent ^{13}C atoms will show carbon-carbon spin-spin coupling that can be revealed by NMR spectroscopy. This method gave interesting information about the folding patterns of the polyketides.[18]

V. PHASES IN THE BIOSYNTHESIS OF POLYKETIDES

It is generally agreed[1,4,5,19] that the biosynthesis of polyketides includes two main phases:

1. Chain assembly and cyclization, where condensation of acetate and malonate units gives a β-polyketoacyl intermediate compound. This phase is supposedly carried out on the surface of an enzyme complex. Cyclization stabilizes the intermediate compound that leaves the enzyme surface by hydrolysis.
2. The secondary transformation phase, where less specific enzymes take over the biosynthetic process. This phase is responsible for the great diversity observed among polyketides. During this phase, the enzyme-free intermediate compound undergoes one or more of the following reactions: methylation, reduction, oxidation, chlorination, or ring cleavage.

A. Chain Assembly and Cyclization

Although many attempts were made to purify and characterize polyketide synthetases, none of these attempts resulted in a completely pure preparation. Purification of 6-methylsalicylic acid synthetase is an outstanding example of these efforts. Dimroth and coworkers[20] obtained that enzyme from *Penicillium patulum* and then purified it 100-fold. Their preparation, however, was not consistent with that prepared by Light.[6]

6-Methylsalicylic acid synthetase from both preparations proved to be a multienzyme complex, as is true for fatty acid synthetase. Also, like fatty acid synthetase, 6-methylsalicylic acid synthetase is inhibited by the sulfhydryl group inhibitors, iodacetamide and *N*-ethylmaleimide[20] The pH-dependent inhibition with *N*-ethylmaleimide and the pH-independent inhibition with iodacetamide imply that 6-methylsalicylic acid synthetase contains two kinds of sulfhydryl groups.

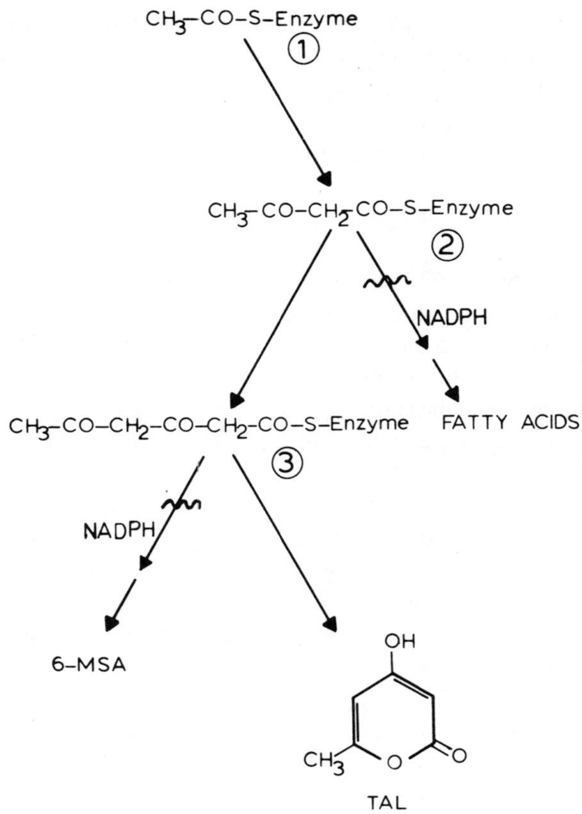

FIGURE 3. Biosynthesis of triacetic acid lactone by fatty acid synthetase and polyketide synthetase in the absence of NADPH. (1) Enzyme-bound acetate; (2) enzyme-bound acetoacetate; (3) enzyme-bound diketoacyl, TAL = triacetic acid lactone, 6-MSA = 6-methylsalicylic acid. ⁓ = biosynthetic block through lack of NADPH.

In the absence of the reduced coenzyme NADPH, 6-methylsalycilic acid synthetase converts acetyl-CoA and malonyl-CoA into triacetic acid lactone[20] (Figure 3). From this experiment, Dimroth and colleagues[20] concluded that reduction of the β-polyketoacyl intermediate (of 6-methylsalicylic acid) occurs at the triacetic acid level and before the final condensation with malonyl-CoA takes place (Figure 3). Working with fatty acid synthetase from yeast, and in the absence of NADPH, Yalpani[21] also obtained the same metabolite (triacetic acid lactone) (Figure 3).

Dimroth et al.[22] demonstrated further similarities between 6-methylsalicylic acid and fatty acid synthetases. Both systems exhibited malonyl-CoA carboxylase activity upon treatment with iodoacetamide. 6-Methylsalicylic acid synthetases had obvious specificity toward acetyl-CoA; when it was replaced with propionyl-CoA or other higher homologue, the rate of reaction dropped significantly. Like fatty acid synthetase, 6-methylsalicylic acid synthetase has acyl transferase activity; the enzyme transacylated the acetyl moiety of acetyl-CoA to pantetheine.

Other polyketide synthetase systems may also be similar to fatty acid synthetases. Sjöland and Gatenbeck[23] found that replacement of acetyl-CoA with propionyl-CoA in the alternariol biosynthesizing system dramatically decreased the reaction rate. Fatty acid synthetase and polyketide synthetase, however, differ in the extent of reductive steps they exert on intermediate compounds. With fatty acid synthetase, every condensation of the malonyl group (i.e., elongation of the chain by a C_2-unit) is accompanied by two reduction steps that require two molecules of NADPH.[24] For comparison, synthesis of 6-methylsalicylic acid requires the condensation of three malonyl units, but only one reduction step is required.[20]

1. Polyketide Synthetase and Chain Assembly

The close analogy between the two systems, fatty acid synthetase and polyketide synthetase, may warrant extrapolating the well-characterized mechanism involved in fatty acid biosynthesis to that of polyketides. The hypothetical scheme that will be presented derives its main features from those proposed by Lynen[25,26] and Bu'Lock[3] with modifications based on more recent findings about polyketide synthetase[27] and yeast fatty acid synthetase.[28]

Analogous to fatty acid synthetase, polyketide synthetase is a multifunctional protein thatgenerally carries out the following enzymatic activities: (1) transacylation of acetyl and malonyl species, (2) condensation of acetyl, β-polyketoacyl, or reduced β-polyketoacyl moieties with the elongating malonyl units, (3) reduction of β-polyketoacyl intermediate compounds, and (4) dehydration of the reduced intermediate compound. Furthermore, the multifunctional protein carries two sulfhydryl groups where acyl groups can be attached: one of a cysteine residue at the site of condensing activity (Enzyme-SH) and the other on a flexible arm of 4′-phosphopantethiene that is directly attached to the protein (Pantethiene-SH). The polyketide synthetase can be viewed as a multifunctional protein of more than one subunit (probably two, analogous with yeast fatty acid synthetase) rather than a multienzyme complex. The enzymatic activity of the system is distributed between these subunits.

The first step of polyketide synthesis is the transfer of acetyl and malonyl moieties from acetyl-CoA and malonyl-CoA to the Enzyme-SH and Pantethiene-SH groups, respectively. That transfer is mediated and catalyzed by the acyl transfer activity of the multifunctional protein. With the help of the flexible arm (about 20 Å in length), the malonyl group is at juxtaposition on the condensation activity site where the acetyl group is attached. At this site, decarboxylative acylation takes place, wherein C_1 of the acetyl groups is covalently linked to C_2 of the malonyl moiety with the release of CO_2 (Figure 4).

The decarboxylative acylation is facilitated by a coordination between a metal ion attached to the active site and oxygen atoms of the carbonyl or carboxyl groups of the reactants. At this point, the acetoacetyl moiety is attached to both the arm and the active site through thiol ester linkage and metal chelation, respectively.

The acetoacetyl moiety is internally transacylated to the Enzyme-SH site, and the Pantethiene-SH site is freed for a new malonyl group. The elongation of the ketoacyl moiety goes on and every time a new malonyl is reacted with the protein-bound ketoacyl intermediate compound, its chain length increases by two carbon units. Metal chelation makes feasible the folding of the β-polyketoacyl intermediate compound to a stable conformation, which dictates the specificity of the cyclization process that follows. It is possible that the geometry and position of the metal ion relative to the active center as well as the geometry of the active site itself determine the stereostructure of the final polyketide. The electronic delocalization that should accompany metal chelation with the enolizable carbonyl groups of the polyketoacyl may ease the electrophilic substitution that occasionally happens at the methylene groups of the polyketide intermediate compound.

Involvement of a metal in the proposed mechanism is supported by some experimental evidence. In their model system, Kobuke and Yoshida[29] noted the importance of magnesium in the condensation of malonate and acetate. With another model system for the biosynthesis of polyketides and fatty acids, Scott et al.[30] found that the role of magnesium was the control of *C*-acetylation over possible *O*-acetylation of acetate derivative. A scheme for chain assembly is shown in Figure 4.

As indicated before in the 6-methylsalicylic acid synthetase system, the absence of NADPH resulted in an unstable intermediate compound that went through lactonization and was

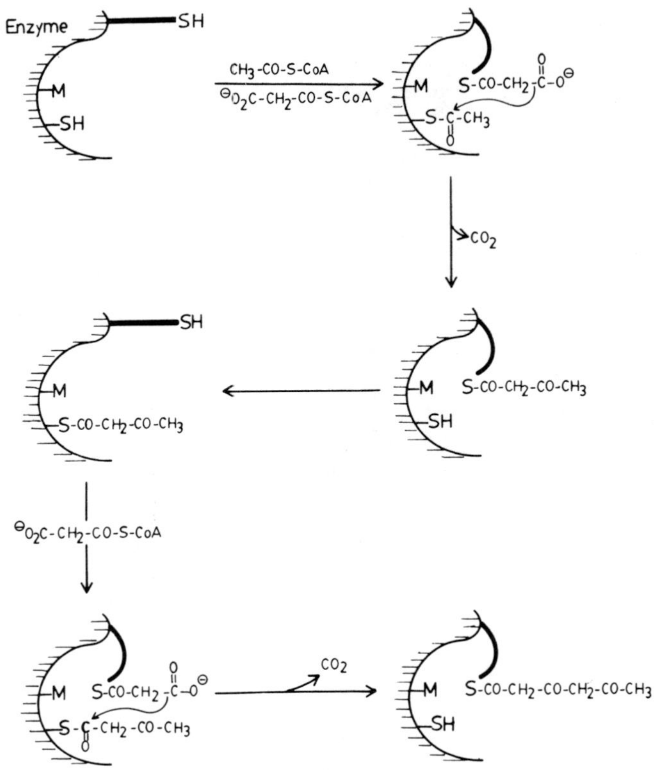

FIGURE 4. Chain assembly on polyketide synthetase. Note the involvement of two –SH groups; one is directly attached to the active site and the other is attached to a flexible arm of 4′-phosphopantethiene. Note also the presence of a metal ion (M) attached to the active center. Metal chelation is not illustrated to simplify the drawing.

released from the enzyme at an early state of chain assembly.[20] This may indicate the importance of the reduction step and probably the subsequent dehydration (giving a double bond) in stabilizing the intermediate compound. Other polyketide synthetases (e.g., that of alternariol biosynthesis), however, do not require NADPH, and the intermediate compound is not reduced. Metal chelation in this instance may play a significant role in stabilizing those intermediate compounds.

B. Secondary Transformation

As discussed earlier, synthesis of the first free polyketide intermediate compound is assumed to take place on the surface of a multifunctional protein. Modification of the intermediate compound is accomplished by several enzymes. Unlike in the previous phase, a greater number of enzymes of secondary transformation were purified and studied. However, confusion about the route that these intermediate compounds follow still exists, and may be greater than that encountered in the previous phase. The lack of absolute specificity of many enzymes of this phase, and failure to define the obligatory intermediate compounds of several steps led to what Bu'Lock described as a "metabolic grid".[31] It simply means that it is possible to demonstrate the incorporation of putative intermediate "E" into a product "C" without "E" being a normal obligatory intermediate compound of the route from A to C:

$$\begin{array}{c} A \rightarrow B \rightarrow C \\ \upharpoonleft\downharpoonright \\ D \rightarrow E \rightarrow F \end{array}$$

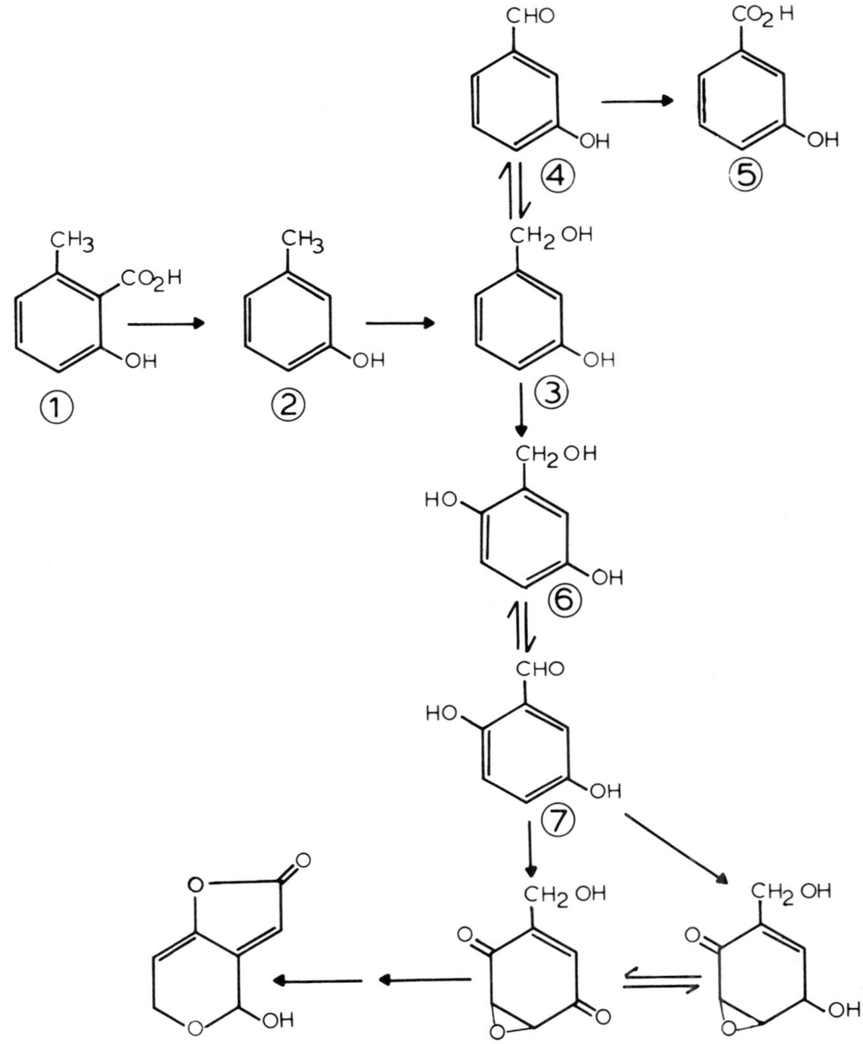

FIGURE 5. Conversion of 6-methylsalicylic acid into patulin. Note the involvement of a metabolic grid. (1) 6-Methylsalicylic acid, (2) *m*-cresol, (3) *m*-hydroxybenzyl alcohol, (4) *m*-hydroxybenzaldhyde, (5) *m*-hydroxybenzoic acid, (6) gantisyl alcohol, (7) gentisaldhyde, (8) phyllostine, (9) isoepoxydon, (10) patulin.[32,59]

The lack of specificity of some of the enzymes can cause the interconnecting of several pathways, giving more than one route for production of a single metabolite. This resulted in ambiguities in many proposed polyketide pathways and led to emphasis on such expressions as ''obligate intermediate compounds'', ''derailments'', and ''co-metabolites''. An example of a metabolic grid (Figure 5) is demonstrated by processes involved in the conversion of 6-methylsalicylic acid into patulin.

In the following paragraphs, we will discuss some of the important reactions of the secondary transformation phase and the enzymes that are involved in the reactions.

1. Reduction and Oxidation

In addition to reduction of the protein-bound β-polyketoacyl derivatives, biosynthesis of some polyketides may include further reduction steps. Oxidation, however, is the most common reaction in the secondary transformation phase. Conversion of 6-methylsalicylic acid into patulin involves at least five enzymes. Three of them are oxygenases and one is a dehydrogenase.[32] The biosynthesis of patulin has been studied more than have other biosynthetic routes in polyketide synthesis. There are indications that oxygenases also are common in the biosynthesis of other polyketides.[33] Oxidative enzymes catalyze such reactions as hydroxylation, oxidative cleavage of rings, and oxidation of alcohol or the aldehyde groups.

There is increasing evidence for the involvement of mixed-function oxidases of the cytochrome P-450, NADPH-cytochrome reductase type in some of these oxidation reactions. Two of the oxidases that catalyze conversion of 6-methylsalicylic acid into patulin are of the former type.[34,35] They catalyze the introduction of an oxygen atom into the substrate with the reduction of the other to water, and they require reduced NADPH for their activity.

$$\text{X-H} + O_2 + \text{NADPH} + H^+ \xrightarrow{\text{Monooxygenase}} \text{X-OH} + H_2O + \text{NADP}^+$$

When these hydroxylation reactions are followed by a dehydration step, an oxygen atom is lost and a double bond is created in the polyketide intermediate compound.

2. Methylation

During the biosynthesis of some polyketides, one or more methylation steps may be required. The methylation may occur on the synthetase-bound intermediate compounds or after the intermediate compound is released. Barber and associates[36] presented evidence to show that all three methylation steps involved in the biosynthesis of citrinin by *Penicillium citrinum* take place on the polyketide synthetase. However, methylation of the synthetase-free intermediate compound is noted in the biosynthesis of some polyketides. For example, the methylation steps in aflatoxin biosynthesis take place on a synthetase-free intermediate compound (probably versicolorin A).[37]

Methyl groups originate from the C_1 pool in mold cells where N^5, N^{10}-methylene tetrahydrofolic acid is involved. The methyl group is transferred to methionine, which is the immediate precursor of methylation reactions. Both *C*-methylation and *O*-methylation of polyketide intermediate compounds were reported.[5]

VI. ORIGIN OF PRECURSORS OF POLYKETIDES

Many researchers have demonstrated the need for acetate and malonate in the biosynthesis of polyketides, and so this biosynthetic route is considered an acetate-polymalonate pathway. Beside these two building blocks, the reduced coenzyme NADPH is commonly required, whereas methionine is less frequently used.

A. Acetyl-CoA

The activated acetate (acetyl-CoA) is the form that participates in polyketide biosynthesis. One molecule of acetyl-CoA is required for each synthesized polyketide molecule. The acetyl-CoA contributes the first two carbon atoms of the polyketide progenitor, with malonyl-CoA contributing the carbon atoms of the rest of the chain. There are several sources of acetyl-CoA in mold cells (Figure 6).

1. Glycolysis

Glycolysis accounts for most of the acetyl-CoA generated in mold cells.[28] In cytosol, glycolysis results in pyruvic acid that is decarboxylated by pyruvic dehydrogenase of the mitochondria, releasing acetyl-CoA in the mitochondrial matrix.[24]

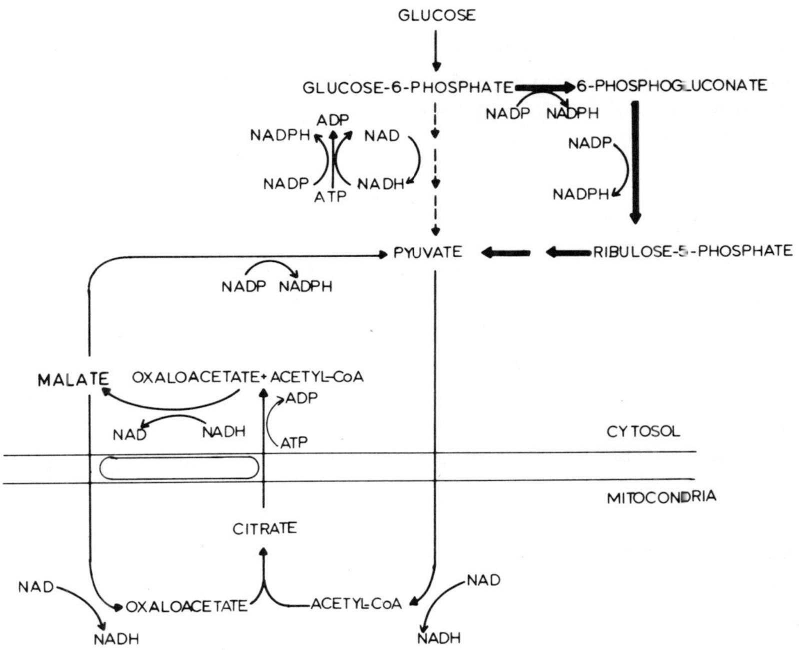

FIGURE 6. Sources of NADPH in the mold cell.

The synthetases of fatty acids[24,28] and, presumably of polyketides[38] are in the cytosol. Acetyl-CoA must be available in the cytosol to satisfy these needs, but it cannot cross the mitochondrial membrane barrier. There are three ways,[28] of unequal likelihood, by which acetyl-Coa is available for fatty acid and polyketide synthesis in the cytosol. (1) Acetyl-CoA enters the tricarboxylic acid cycle where it combines with an oxaloacetate molecule to form citrate. The latter can cross the mitochondrial membrane where it is broken down into oxaloacetate and acetyl-CoA in the cytosol by ATP citrate lyase. (2) The acetyl moiety of acetyl-CoA may be transported across the mitochondrial membrane, facilitated by carnitin. Acetyl-CoA is then regenerated in the cytosol. (3) Acetyl-CoA may hydrolyze in the mitochondrial matrix, and acetate diffuses out of the membrane where acetyl-CoA can be easily formed by acetate thiokinase in the cytosol. It seems that the first route is the most likely one.[24]

2. *Degradation of Fatty Acids*

β-Oxidation of fatty acids in the mitochondria generates acetyl-CoA, which may be transported to the cytosol by the previous mechanisms.

3. *Other Sources*

Acetate in the medium of the mold enters the cell where it is converted to acetyl-CoA in the cytosol by acetic thiokinase.[28] Degradation of amino acids contributes to the acetyl-CoA pool in mold cells.

Findings consistent with this pattern of acetyl-CoA compartmentalization were obtained during a study on aflatoxin biosynthesis.[38] The study indicated that label from acetate was incorporated into aflatoxin faster than that from glucose. The study also showed that aflatoxin was labeled exclusively by acetate until most of the exogenous acetate was exhausted, and then carbon atoms from glucose started to be incorporated into the toxin.

B. Malonyl-CoA

Malonyl-CoA is formed in molds from acetyl-CoA by carboxylation. The enzyme that catalyzes this reaction (acetyl-CoA carboxylase) contains biotin, and the reaction requires bicarbonate. The enzyme is specific for Mn^{2+}. The reaction has two main steps: (1) a carboxylation reaction involving formation of a carboxybiotin-enzyme intermediate compound and (2) transcarboxylation reaction whereby acetyl-CoA is converted into malonyl-CoA.[24,28]

$$HCO_3 + ATP + Enzyme \rightarrow Enzyme\text{-}COO^- + ADP + Pi$$

$$Enzyme\text{-}COO^- + Acetyl\text{-}CoA \rightarrow Malonyl\text{-}CoA + Enzyme$$

Carboxylation of acetyl-CoA is commonly considered a rate-limiting step in the biosynthesis of fatty acids.[39] Acetyl-CoA carboxylase is the site for a number of control mechanisms related to the biosynthesis of fatty acids and probably also polyketides. The enzyme is activated by the tricarboxylic acid intermediate compounds, citrate and isocitrate, through an allosteric control.[40] The activity of acetyl-CoA carboxylase is also controlled by feedback inhibition exerted by end products.[41]

C. Nicotinamide Adenine Dinucleotide Phosphate Coenzyme

The reduced coenzyme NADPH is required for reduction of enzyme-bound polyketide intermediate compounds. Reactions involving monooxygenases also require reduced NADPH (see Section V.B). Dehydrogenases involved in polyketide biosynthesis require NADPH for their activity. The oxidation state of the coenzyme in this instance depends on the direction of the reactions.[42] There are three sources of NADPH in growing fungal cells (Figure 6).

1. Catabolism of Glucose in the Hexose Monophosphate (HMP) Pathway

For each CO_2 molecule produced by this route, two molecules of the coenzyme are reduced. This source is significant since molds use the HMP pathway in their catabolism of carbohydrates.[43]

2. Tricarboxylic Acid Intermediate Compounds

Citrate leaves the mitochondria (in exchange with malate) where is is cleaved in the cytosol by ATP-citrate lyase into acetyl-CoA and oxaloacetate. The latter is reduced to malate that goes into the mitochondria as part of the citrate shuttle system. Malate in the cytosol may also be decarboxylated into pyruvate with generation of a molecule of reduced NADPH.[24,44] Isocitrate also may be transported out of mitochondria in exchange for malate.[44] In the cytosol, isocitrate is oxidized (by cytoplasmic isocitrate dehydrogenase) to α-ketoglutarate with the concomitant generation of NADPH.

3. Catabolism of Glucose by the Embden-Meyerhof (EM) Pathway

The catabolism of glucose by the EM pathway gives NADH, which can be used to reduce $NADP^+$.[24] Hult and Gatenbeck[45] proposed a mannitol cycle that generates NADPH from NADH in the polyketide-producing mold *Alternaria alternata*. The net result of the cycle is as follows:

$$NADH + NADP^+ + ATP \rightarrow NAD^+ + NADPH + ADP + Pi$$

Only a few polyketides are synthesized without reduction of the enzyme-bound intermediate compounds. Such exceptions include orsellinic acid and alternariol. Beside the common substrates we dealt with so far, there are others that may contribute to the structure of polyketides. The methyl group from the C_1-pool is included in many polyketides. Larger units, obtained from other biosynthetic routes, may combine with a polyketide intermediate compound. For example, phenylalanine (from the shikimic acid pathway) combines with a polyketide intermediate compound to form ochratoxin A.[46]

VII. REGULATION OF POLYKETIDE BIOSYNTHESIS

Biosynthesis of polyketides and other secondary metabolites is an outcome of a variety of primary and secondary metabolic processes. The regulation of each of these numerous reactions can theoretically control, to various degrees, the amount of the final product. Unlike primary metabolism, regulation of secondary metabolism, in general, is inherently complicated and lacks the clear-cut experimental evidence available for primary metabolism. A reason for this, beside the one just given, is that most polyketide pathways have not been studied at the enzymatic level. This leaves us in doubt about the proper route these pathways follow, and there may be myths about the regulatory mechanisms that control them. In spite of these limitations and difficulties, attempts were made to shed light on this problem.[3,8,47] In this section, we will make our own attempt at this without forgetting that most of what has been reported earlier was speculative. This attempt will focus on the peculiar features of polyketides.

A. Growth-Related Regulation

There are two schools of thought about the relationship between growth and secondary metabolism. The first considers that phases of growth are responses to the environment and so is the production of a secondary metabolite. The other believes in the "commitment" of mold cells to differentiate and produce secondary metabolites. The first view was based on the findings of Borrow et al.,[7] Taber,[48] and Bu'Lock and associates.[49] The approach was refined by Bu'Lock[3,8,47] and adopted by many researchers working with polyketides. In a well-phase submerged culture of mold, Bu'Lock identified two phases of growth: trophophase and idiophase. During trophosphase, nutrients are taken up in a balanced ratio and actual replicatory growth is observed. Once a nutrient (other than carbon source) becomes limited, growth enters the idiophase. In this phase, replicatory growth slows down or completely stops (though dry weight may continue to increase) and biosynthesis of secondary metabolites commences. During the idiophase there is a series of metabolic events that leads to production of secondary metabolites. The series of events that leads to production of aflatoxin (and probably other polyketides), as viewed by Maggon et al.,[19] starts with the exhaustion of nitrogen or phosphorus in the medium at the end of trophophase. Syntheses of nucleic acids and proteins and, consequently, replicatory growth are inhibited, and pyruvic acid and other α-ketoacids accumulate. At the commencement of idiophase, the enzymes of aflatoxin biosynthesis are derepressed and the tricarboxylic acid cycle activity reaches a maximum. As the levels of pyruvate and other intermediate compounds increase to toxic levels, cell replication stops and mold cells shift these intermediate compounds to secondary metabolism such as synthesis of aflatoxins and lipids. A sharp division of growth into well-defined phases and complete independence of secondary metabolism from growth is not frequently met in mold cultures,[47] particularly in quiescently incubated cultures.[9] Bu'Lock[8] admitted that growth and secondary metabolism are not mutually exclusive and incompatible activities.

In the growing mycelium, hyphal tips are always younger and less differentiated than the more distal portions.[50] Accordingly, in a growing mold culture, there should be an age distribution of mycelium, and older parts of the mycelium should be metabolically different from younger ones. Since cells of different ages coexist in a mold culture, independence of secondary metabolism from growth is not likely. This kind of logic led to the evolution of a second school of thought, which believes that production of secondary metabolites is a function of cell age.[51,52]

In their attempt to put the relationship between growth and secondary metabolism in simple terms, and after reviewing previous ideas,[53] Brown and Vass[52] developed a kinetic model to describe that relationship. The model depends on the hypothesis of cell maturation, i.e., newly formed mold cells need time (t_m) to mature and become capable of producing secondary metabolites. Consequently, the rate of biosynthesis of a secondary metabolite after (t) hours of incubation shoud depend on the rate of cell maturation, which in turn equals the rate of cell formation (rate of growth) that occurred in the culture ($t - t_m$) hours earlier. This relationship was mathematically expressed as follows:

$$\left(\frac{dY}{dt}\right)_t = \alpha \left(\frac{dX}{dt}\right)_{t-t_m}$$

where dY/dt is the rate of product formation, dX/dt is growth rate, and α is the product formation rate constant.

The model implies that secondary biosynthesis is an echo of replicatory growth. It also leaves the door open for speculation on which environmental factors might be significant in regulating secondary metabolism by altering the maturation time (t_m) and degree of dependence of secondary metabolism on previous growth (α).

The model was useful in explaining the growth-secondary metabolism relationship during production of macrolide antibiotics by *Streptomyces* sp.[54] and production of aflatoxins as regulated growth and biosynthetic inhibitors.[10,11] Martin and McDaniel[54] viewed maturation time (t_m) as a period of transition between the trophophase and idiophase.

Aging and differentiation, according to Bu'Lock,[8] like production of secondary metabolites, are provoked by environmental factors. Thus he believes such modeling could be biologically misleading.

B. Regulation through Primary Metabolism

Although this topic can be partly covered under growth-related regulation, we preferred to deal with it separately. Some primary metabolites share precursors in common with polyketides. Acetyl-CoA may be taken up by the tricarboxylic acid cycle, and its intermediate compounds can be channelled into primary metabolites. NADPH is needed for many primary anabolic processes in the cell. Accordingly, regulating these primary processes theoretically can affect the biosynthesis of polyketides. An example of this kind of regulation is given by Shih and Marth[55] in their studies on aflatoxins. Treatment of a mold culture with a sublethal dose of sodium azide enhanced production of aflatoxins. Sodium azide is an inhibitor of terminal electron transfer. The authors speculated that the decrease of respiratory activity in the presence of that inhibitor resulted in accumulation of acetate, the precursor for aflatoxin. Furthermore, inhibition of terminal electron transport will inhibit oxidation of reduced nicotinamide dinucleotide that is needed for biosynthesis of aflatoxins. Another example is provided by a study on specific inhibition of a TCA cycle-related intermediate compound and its effect on biosynthesis of alternariol and fatty acid.[44] (−)Hydroxycitrate is a powerful inhibitor of ATP-citrate lyase, which makes acetyl-CoA available for the synthesis of polyketides and fatty acids (Figure 6). Inhibition of this enzyme led to a decrease in biosynthesis of fatty acids, but not alternariol, from exogenously supplied acetate. Since the biosynthesis of alternariol does not require NADPH, but the biosynthesis of fat does, the investigators believe that the former pattern of inhibition resulted from a lack of reduced NADPH (see other interpretation in later section). Note that by the action of citrate lyase, citrate in the cytosol gives oxaloacetate that is further transformed to malate. The latter is oxidized to pyruvate with production of NADPH (Figure 6). The authors also noted the reversal of inhibition by malate. Biosynthesis of alternariol probably is a good model for studies on the regulation of polyketides as this product does not require NADPH or further secondary transformations.[33]

Generally speaking, the biosynthesis of polyketides and related products is less NADPH-demanding than is that of fatty acids.[56] Furthermore, some researchers suggested that production of polyketides reflects a limiting availability of NADPH for fatty acid biosynthesis.[45,57]

C. Direct Regulation in Secondary Metabolism

It is generally believed that enzymes of the polyketide route are inducible rather than constitutive. Enzymes that catalyze transformation of 6-methylsalicylic acid into patulin appear in succession in mold cells.[12,34] This gives an indication that these enzymes may be sequentially induced.[47] Induction could be affected by primary or secondary metabolic factors. Some of the enzymes that catalyze secondary transformation, in the aflatoxin biosynthetic pathway, are believed to be induced by the appearance of an enzyme-free anthraquinone precursor.[58]

Carboxylation of acetyl-CoA to generate chain-elongating units of malonate is a limiting step in fatty acid biosynthesis.[28] If that is true for the biosynthesis of polyketides, regulation of that important step (see Section VI) could have a pronounced effect on the biosynthesis of polyketides. This kind of regulation is hinted at in a study on aflatoxin. At certain levels of citrate, biosynthesis of aflatoxin by lysates of spheroplasts from *Asperigillus parasiticus* was enhanced. Higher or lower levels of citrate, however, were inhibitory.[17]

It is likely that the availability of polyketide synthetase (at least at certain stages of growth) is more limiting than the availability of its substrates. Affinity of polyketide synthetase, relative to that of fatty acid synthetase, for common substrates is another factor that may be of significance. A result from a study of alternariol may be explained by this postulate. Hult and Gatenbeck[44] found that an inhibitor of ATP-citrate lyase, which affects the availability of acetyl-CoA (and NADPH), decreased incorporation of exogenously added acetate into fatty acids but not into alternariol.

As we indicated before, many of the comments in this last section are speculative and are not a substitute for the serious investigations that are needed to improve our understanding of the regulation of polyketide biosynthesis.

ACKNOWLEDGMENTS

This chapter is a contribution from the College of Agricultural and Life Sciences, University of Wisconsin-Madison, Madison, Wisc. Preparation of the manuscript was supported in part by the Dairy Research Foundation of the National Dairy Council, Rosemont, Ill. and by Kraft, Inc., Glenview, Ill.

REFERENCES

1. **Money, T.,** Biosynthesis of polyketides, in *Biosynthesis, Specialist Periodical Reports,* Vol. 2, Geissman, T. A., Ed., The Chemical Society, London, 1973, 183.
2. **Simpson, T. J.,** Biosynthesis of polyketides, in *Biosynthesis, Specialist Periodical Reports,* Vol. 5, Bu'Lock, J. D., Ed., The Chemical Society, London, 1977, 1.
3. **Bu'Lock, J. D.,** *Essays in Biosynthesis and Microbial Development,* John Wiley & Sons, New York, 1967, chaps. 2 and 3.
4. **Turner, W. B., Ed.,** *Fungal Metabolites,* Academic Press, New York, 1971, chap. 5.
5. **Turner, W. B.,** Polyketides and related metabolites, in *The Filamentous Fungi,* Vol. 2, Smith, J. E. and Berry, D. R., Eds., John Wiley & Sons, New York, 1976, chap. 15.
6. **Light, R. J.,** Enzymatic studies on the polyketide hypothesis, *J. Agric. Food Chem.,* 18, 260, 1970.
7. **Borrow, A., Jefferys, E. G., Kessell, R. H. J., Lloyd, E. C., Lloyd, P. B., and Nixon, I. S.,** The metabolism of *Gibberella fujikuroi* in stirred culture, *Can. J. Microbiol.,* 7, 227, 1961.

8. **Bu'Lock, J. D.,** Secondary metabolism in fungi and its relationships to growth and development, in *The Filamentous Fungi,* Vol. 1, Smith, J. E. and Berry, D. R., Eds., John Wiley & Sons, New York, 1975, chap. 3.
9. **Yousef, A. E. and Marth, E. H.,** Growth and synthesis of aflatoxin by *Aspergillus parasiticus* in the presence of sorbic acid, *J. Food Prot.,* 44, 736, 1981.
10. **Yousef, A. E. and Marth, E. H.,** Kenetics of aflatoxin biosynthesis by *Aspergillus parasiticus* in the presence of N α-palmitoyl-L-lysyl-L-lysine ethyl ester dihydrochloride or dichlorvos, *Biotechnol. Bioeng.,* 25, 671, 1983.
11. **Yousef, A. E. and Marth, E. H.,** Kinetics of growth and accumulation of aflatoxin B_1 by *Aspergillus parasiticus* in the presence of butylated hydroxyanisol, isoprothiolane, and nystatin, *Biotechnol. Bioeng.,* 26, 6, 1984.
12. **Forrester, P. I. and Gaucher, G. M.,** Conversion of 6-methylsalicylic acid into patulin by *Pencillium urticae, Biochemistry,* 11, 1102, 1972.
13. **Shantha, T. and Murthy, V. S.,** Influence of tricarboxylic acid cycle intermediates and related metabolites on the biosynthesis of aflatoxin by resting cells of *Aspergillus flavus, Appl. Environ. Microbiol.,* 42, 758, 1981.
14. **Yousef, A. E. and Marth, E. H.,** Incorporation of [^{14}C]acetate into aflatoxin by resting cultures of *Aspergillus parasiticus* in the presence of antifungal agents, *Eur. J. Appl. Microbiol. Biotechnol.,* 18, 103, 1983.
15. **Singh, R. and Hsieh, D. P. H.,** Aflatoxin biosynthetic pathway: elucidation by using blocked mutants of *Aspergillus parasiticus, Arch. Biochem. Biophys.,* 178, 285, 1977.
16. **Dutton, M. F. and Anderson, M. S.,** The use of fungal protoplasts in the study of aflatoxin biosynthesis, *Experientia,* 15, 22, 1978.
17. **Tyagi, J. S., Tyagi, A. K., and Venkitasubramanian, T. A.,** Some factors regulating [1-^{14}C]acetate incorporation into aflatoxins by spheroplasts and spheroplast lysates from *Aspergillus parasiticus, Toxicon,* 19, 445, 1981.
18. **Seto, H., Cary, L. W., and Tanabe, M.,** Utilization of carbon-13-carbon-13 coupling in structural and biosynthetic studies. Fourier transform carbon-13 nuclear magentic resonance spectrum of mollisin, *J. Chem. Soc. Chem. Commun.,* 22, 867, 1973.
19. **Maggon, K. K., Gupta, S. K., and Venkitasubramanian, T. A.,** Biosynthesis of aflatoxins, *Bacteriol. Rev.,* 41, 822, 1977.
20. **Dimroth, P., Walter, H., and Lynen, F.,** Biosynthesis of 6-methylsalicylic acid, *Eur. J. Biochem.,* 13, 98, 1970.
21. **Yalpani, M., Willecke, K., and Lynen, F.,** Triacetic acid lactone, a derailment product of fatty acid biosynthesis, *Eur. J. Biochem.,* 8, 495, 1969.
22. **Dimroth, P., Ringelmann, E., and Lynen, F.,** 6-Methylsalicylic acid synthetase from *Penicillium patulum,* some catalytic properties of the enzyme and its relation to fatty acid synthetase, *Eur. J. Biochem.,* 68, 591, 1976.
23. **Sjöland, S. and Gatenbeck, S.,** Studies on the enzyme synthesizing the aromatic product alternariol, *Acta Chem. Scand.,* 20, 1053, 1966.
24. **White, A., Handler, P., Smith, E. L., Hill, R. L., and Lehman, I. R.,** *Principles of Biochemistry,* McGraw-Hill, New York, 1978, chaps. 12, 14, and 17.
25. **Lynen, F.,** Biosynthetic pathways from acetate to natural products, *Pure Appl. Chem.,* 14, 137, 1967.
26. **Lynen, F.,** Multienzyme complex of fatty acid synthetase, in *Organizational Biosynthesis,* Vogel, H. J., Lampen, J. O., and Bryson, V., Eds., Academic Press, New York, 1967, 243.
27. **Dimroth, P., Greull, G., Seyffert, R., and Lynen, F.,** 6-Methylsalicylic acid synthetase, *Hoppe-Seyler's Z. Physiol. Chem.,* 353, 126, 1972.
28. **Weete, J. D.,** *Lipid Biochemistry of Fungi and Other Organisms,* Plenum Press, New York, 1980, chap. 4.
29. **Kobuke, Y. and Yoshida, J.,** Decarboxylative acylation of thiolmalonate. A model for the biosynthesis of fatty acids and polyketides, *Tetrahedron Lett.,* 4, 367, 1978.
30. **Scott, A. I., Wiesner, C. J., Yoo, S., and Chung, S. K.,** Acetyl transfer reaction in catechol acetate malonate. A model for biosynthesis of polyketides and fatty acids, *J. Am. Chem. Soc.,* 97, 6277, 1975.
31. **Bu'Lock, J. D.,** *The Biosynthesis of Natural Products,* McGraw-Hill, London, 1965, chap. 3.
32. **Zamir, L. O.,** The biosynthesis of patulin and peniciliic acid, in *The Biosynthesis of Mycotoxins,* Steyn, P. S., Ed., Academic Press, New York, 1980, chap. 7.
33. **Steyn, P. S., Vleggaar, R., and Wessels, P. L.,** The biosynthesis of aflatoxin and its congeners, in *The Biosynthesis of Mycotoxins,* Steyn, P. S., Ed., Academic Press, New York, 1980, chap. 4.
34. **Murphy, G., Vogel, G., Krippahl, G., and Lynen, F.,** Patulin bisoynthesis: the role of mixed-function oxidases in the hydroxylation of m-cresol, *Eur. J. Biochem.,* 49, 443, 1974.
35. **Murphy, G. and Lynen, F.,** Patulin biosynthesis: the metabolism of *m*-hydroxybenzyl alcohol and *m*-hydroxybenzaldhyde by particulate preparations from *Penicillum patulum, Eur. J. Biochem.,* 58, 467,1975.

36. **Barber, J., Carter, R. H., Garson, M. J., and Staunton, J.,** The biosynthesis of citrinin by *Penicillium citrinum, J. Chem. Soc. Perkin Trans. I,* 2577, 1981.
37. **Gorst-Allman, C. P., Steyn, P. S., Wessels, P. L., and Scott, De B.,** Carbon-13 nuclear magentic resonance assignment and biosynthesis of versicolorin A in *Aspergillus parasiticus, J. Chem. Soc. Perkin Trans. I,* 961, 1978.
38. **Hsieh, D. P. H. and Mateles, R. I.,** The relative contribution of acetate and glucose to aflatoxin biosynthesis, *Biochim. Biophys. Acta,* 208, 482, 1970.
39. **Vagelos, P. R.,** Lipid metabolism, *Ann. Rev. Biochem.,* 33, 139, 1964.
40. **Rasmussen, R. K. and Klein, H. P.,** Regulation of acetyl CoA carboxylase of *Saccharomyces cervisiae, Biochem. Biophys. Res. Commun.,* 28, 415, 1967.
41. **Lust, G. and Lynen, F.,** The inhibition of the fatty acid synthetase multienzyme complex of yeast by long-chain acyl coenzyme A compounds, *Eur. J. Biochem.,* 7, 68, 1968.
42. **Scott, A. I. and Beadling, L.,** Biosynthesis of patulin, dehydrogenase and dioxygenase enzymes of *Penicillium patulum, Biol. Chem.,* 3, 281, 1974.
43. **Cochrane, V. W.,** Glycolysis, in *The Filamentous Fungi,* Vol. 2, Smith, J. E. and Berry, D. R., Eds., John Wiley & Sons, New York, 1976, chap. 3.
44. **Hult, K. and Gatenbeck, S.,** Regulation and metabolic background of polyketide formation. I. Effects of (−)-hydroxycitrate and metabolic roles of citrate and molate in fatty acid and polyketide formation, *Acta Chem. Scand.,* B30, 283, 1976.
45. **Hult, K. and Gatenbeck, S.,** Production of NADPH in the mannitol cycle and its relation to polyketide formation in *Alternaria alternata, Eur. J. Biochem.,* 88, 607, 1978.
46. **Maebayashi, Y., Miyaki, K., and Yamazaki, M.,** Application of ^{13}C-NMR to the biosynthetic investigations. I. Biosynthesis of ochratoxin A, *Chem. Pharm. Bull.,* 20, 2172, 1972.
47. **Bu'Lock, J. D.,** Mycotoxins as secondary metabolites, in *The Biosynthesis of Mycotoxins,* Steyn, P. S., Ed., Academic Press, New York, 1980, chap. 1.
48. **Taber, W. A.,** Sequential formation and accumulation of primary and secondary shunt metabolic products in *Claviceps purpurea, Appl. Microbiol.,* 12, 321, 1964.
49. **Bu'Lock, J. D., Hamilton, D., Hulme, M. A., Powell, A. J., Smalley, H. M., Shepherd, D., and Smith, G. N.,** Metabolic development and secondary biosynthesis in *Penicillium urticae, Can. J. Microbiol.,* 11, 765, 1965.
50. **Zalokar, M.,** Growth and differentiation of *Neurospora hyphae, Am. J. Botany,* 46, 602, 1959.
51. **Megee, R. D., III, Kinoshita, S., Fredrickson, A. G., and Tsuchiya, H. M.,** Differentiation and product formation in molds, *Biotechnol. Bioeng.,* 12, 771, 1970.
52. **Brown, D. E. and Vass, R. C.,** Maturity and product formation in cultures of microorganisms, *Biotechnol. Bioeng.,* 15, 321, 1973.
53. **Blanch, H. W. and Rogers, P. L.,** Production of gramicidin S in batch and continuous culture, *Biotechnol. Bioeng.,* 13, 843, 1971.
54. **Martin, J. F. and McDaniel, L. E.,** Kinetics of biosynthesis of polyene macrolide antibiotics in batch culture: cell maturation time, *Biotechnol. Bioeng.,* 17, 925, 1975.
55. **Shih, C. N. and Marth, E. H.,** Aflatoxin formation, lipid synthesis, and glucose metabolism by *Aspergillus parasiticus* during incubation with and without agitation, *Biochim. Biophys. Acta,* 338, 286, 1974.
56. **Miller, P. A., Hash, J. H., Lincks, M., and Bohonos, N.,** Biosynthesis of 5-hydroxytetracycline, *Biochem. Biophys. Res. Commun.,* 18, 325, 1965.
57. **Gatenbeck, S. and Hermodsson, S.,** Enzymatic synthesis of the aromatic product alternariol, *Acta Chem. Scand.,* 19, 65, 1965.
58. **Lin, M. T. and Hsieh, D. P. H.,** Averufin in the biosynthesis of aflatoxin B_1, *J. Am. Chem. Soc.,* 95, 1668, 1973.
59. **Sekiguchi, J. and Gaucher, G. M.,** Patulin bisoynthesis, the metabolism of phyllostine and isoepoxydon by cell-free preparation from *Penicillium urticae, Can. J. Microbiol.,* 25, 881, 1979.

Chapter 2

TOXIC SECONDARY METABOLITES OF *FUSARIUM* SPECIES

Kenji Ishii and Yoshio Ueno

TABLE OF CONTENTS

I. INTRODUCTION

Fungi of the genus *Fusarium* are widely distributed in our environment. They are soil- or air-borne fungi and some are pathogenic to plants. The fungi of *Fusarium* produce a variety of biologically active metabolites. Among the biologically active secondary metabolites, trichothecenes comprise the largest group of compounds. The second largest group of biologically active metabolites is made up of zearalenone and its related compounds. This group of compounds possesses estrogenic activity. In addition to the two groups of biologically active metabolites of *Fusarium* sp., the authors describe butenolide and moniliformin.

II. TRICHOTHECENES

Trichothecenes are produced by various species of fungi of the genera *Fusarium, Trichothecium, Trichoderma, Myrothecium, Cephalosporium, Staschybotrys, Verticimonosporium,* and *Cylindocarpon*. The name "trichothecane" was proposed for the structure of a tricyclic skeleton composed of cyclohexane, cyclopentane, a 6-membered oxyrane ring, and two methyl groups.[1] This name was derived from "trichothecin", which had been the first trichothecene isolated from *Trichothecium roseum*. Naturally occurring trichothecenes have a double bond at C-9 and C-10 and an epoxide group at C-12 and C-13. Trichothecenes are divided into two groups depending on the presence of a macrocyclic ring linking at C-4 and C-15 with di- or triester, namely macrocylcic and non-macrocyclic trichothecenes. Non-macrocyclic trichothecenes are divided into three subgroups (types A, B, and C) depending on the substituent at C-8.[2] The first subgroup (type A) includes trichothecenes having a hydroxyl group or its acyl ester and trichothecenes without a substituent at C-8. The B type of trichothecenes has a carbonyl group at C-8. The C type have an epoxide group at C-7 and C-8, such as crotocin from *Cephalosporium crotocigenum*. Trichothecenes so far isolated from *Fusarium* sp. are members of the A and B type. In Tables 1 and 2, trichothecenes produced by fungi of *Fusarium* are listed. The first trichothecene to be isolated from *Fusarium* fungi was diacetoxyscirpenol.[3] This compound was isolated from *F. scirpi,* along with two minor metabolites, as phytotoxic metabolites. Some trichothecenes from fungi of genera other than *Fusarium* were isolated as antifungal agents. In the case of the genus *Fusarium,* most trichothecenes have been isolated as food contaminants, namely mycotoxins. Chemically, trichothecenes are a sequiterpenoid. The biosynthesis of the skeleton has been postulated to include cyclization of farnesyl pyrophosphate, migration of methyl groups and hydrogen, and formation of trichodiene and trichodiol.

A. Lethal Toxicity

Many studies have been conducted on the acute lethal toxicities of trichothecenes. In Table 3 LD_{50} values of trichothecenes of *Fusarium* sp. are summarized. A variety of animals such as mice, rats, guinea pigs, rabbits, cats, dogs, swine, ducklings, and chicks have been used for the LD_{50} studies. In these studies, the route of administration, the solvents used for dissolving the toxins, and the sex, age, and strain of animals are all different. Therefore, evaluation of the lethal toxicity of toxins is not conclusive, but relative. However, some important information concerning the lethal toxicity of trichothecenes can be obtained. Mice have been most widely used for LD_{50} studies of trichothecenes, and this enables us to compare the toxicities of each toxin. LD_{50} values of intraperitoneally administered toxin in mice range from 3.4 (fusarenon-X) to 70 mg/kg (deoxynivalenol), indicating that the latter toxin is 20 times less toxic than the former one. There is no clear relationship between the structure and toxicity of trichothecenes, but the toxicity is somewhat related to the number of acyl groups in the molecule. For example, the i.p. LD_{50} in mice of T-2 toxin, HT-2 toxin, neosolaniol, and NT-1 are 5.2, 9.0, 14.5, and 28.5 mg/kg, respectively. In addition,

Table 1
TRICHOTHECENES OF *FUSARIUM* SPECIES (A TYPE)

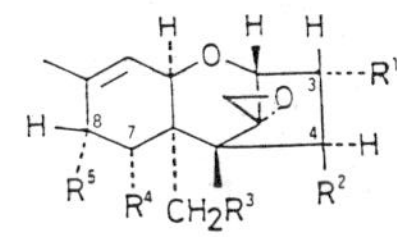

Compounds	R^1	R^2	R^3	R^4	R^5
Scirpentriol	OH	OH	OH	H	H
Monoacetoxyscirpenol	OH	OH	$OCOCH_3$	H	H
4-Acetoxyscirpen-3,15-diol	OH	$OCOCH_3$	OH	H	H
Diacetoxyscirpenol	OH	$OCOCH_3$	$OCOCH_3$	H	H
4,15-Diacetoxytrichothecene-3,7-diol	OH	$OCOCH_3$	$OCOCH_3$	OH	H
4,15-Diacetoxytrichothecene-3,7,8-triol	OH	$OCOCH_3$	$OCOCH_3$	OH	OH
4,8,15-Triacetoxytrichothecene-3,7-diol	OH	$OCOCH_3$	$OCOCH_3$	OH	$OCOCH_3$
T-2 tetraol	OH	OH	OH	H	OH
15-Acetyl T-2 tetraol	OH	OH	$OCOCH_3$	H	OH
3′-Hydroxy T-2 triol	OH	OH	OH	H	$OCOCH_2COH(CH_3)_2$
NT-2	OH	$OCOCH_3$	OH	H	OH
NT-1	OH	$OCOCH_3$	OH	H	$OCOCH_3$
Neosolaniol	OH	$OCOCH_3$	$OCOCH_3$	H	OH
8-Acetylneosolaniol	OH	$OCOCH_3$	$OCOCH_3$	H	$OCOCH_3$
HT-2 Toxin	OH	OH	$OCOCH_3$	H	$OCOCH_2CH(CH_3)_2$
3′-Hydroxy HT-2 toxin	OH	OH	$OCOCH_3$	H	$OCOCH_2COH(CH_3)_2$
T-2 toxin	OH	$OCOCH_3$	$OCOCH_3$	H	$OCOCH_2CH(CH_3)_2$
Acetyl T-2 toxin	OCOCH	$OCOCH_3$	$OCOCH_3$	H	$OCOCH_2CH(CH_3)_2$
Calonectrin	OCOCH	H	$OCOCH_3$	H	H
Deacetlcalonectrin	OCOCH	H	OH	H	H

Table 2
TRICHOTHECENES OF *FUSARIUM* SPECIES (B TYPE)

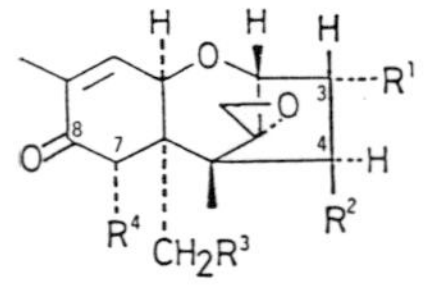

Compounds	R^1	R^2	R^3	R^4
Nivalenol	OH	OH	OH	OH
Fusarenone-X	OH	$OCOCH_3$	OH	OH
Diacetylnivalenol	OH	$OCOCH_3$	$OCOCH_3$	OH
Deoxynivalenol	OH	H	OH	OH
3-Acetyldeoxynivalenol	$OCOCH_3$	H	OH	OH
15-Acetyldeoxynivalenol	OH	H	$OCOCH_3$	OH
3,15-Diacetyldeoxynivalenol	$OCOCH_3$	H	$OCOCH_3$	OH
Trichothecene-3,4,5-triol-8-one	OH	OH	OH	H

compounds lacking a hydroxyl group at C-4 in B-type trichothecenes, such as deoxynivalenol and its acetate, are less toxic than others, suggesting that oxygen at C-4 is important for toxicity.

As for the administration route, the LD_{50} values of fusarenon-X in mice are 3.4 (i.v.), 3.4 (i.p.), 4.2 (s.c.), and 4.5 (p.o.) mg/kg, and this may suggest that the acute toxicity of trichothecene is not markedly different when administered by different routes. In the case of deoxynivalenol, the i.p. LD_{50} is 70 mg/kg and the p.o. LD_{50} is 46 mg/kg. In addition, the ratio of the maximum to the minimum lethal dose is approximately 2, namely the dose-response curve of the lethal toxicity is very sharp. These facts suggest that toxins are efficiently absorbed into the body irrespective of the route of administration.

T-2 toxin has been studied in several species of animals, and the p.o. LD_{50} (mg/kg) is 10.5 in mice, 5.2 in rats, 3.06 in guinea pigs, and 6.1 in trout. These data indicate that the LD_{50} of T-2 toxin in the different species are comparable. Newborn animals are highly sensitive to trichothecenes. For example, the s.c. LD_{50} in mice of T-2 toxin, diacetoxyscirpenol, and fusarenon-X are 0.15, 0.17, and 0.2 mg/kg, respectively, while those in 6-week-old mice are 5.2, 2.3, and 3.4 mg/kg, respectively.

In animals administered with the trichothecenes, toxicity was primarily characterized by effects on the epithelial mucosa of the stomach and small and large intestine, resulting in mucosal erosion and hemorrhagic lesions; radiomimetic cellular injuries, such as karyorrhexis and cellular damage, were observed in the actively dividing cells of thymus, spleen, ovary, testis, and lymph node.[4]

B. General Pharmacological Effect

After the administration of sublethal dose of trichothecenes, mice and rats show symptoms of moderate posterior paresis, loss of tension in skeletal muscle, and diarrhea.[5] In other animals, such as cats, dogs, and ducklings, emesis and vomiting are induced shortly after administration of toxins and this vomiting occurs frequently with intervals.

Matsuoka et al.[6] investigated the general pharmacological effects of fusarenon-X in mice and rats. This mycotoxin induced hypothermia but did not cause appreciable behavioral changes in mice. In rats anesthetized with ether, this toxin evoked a rise in blood pressure and a decrease in the respiratory rate, but did not induce a significant change in the heart rate. No detectable effects on spontaneous activity and responsiveness to agonists were

Table 3
LD_{50} VALUES (MG/KG) OF TRICHOTHECENES

Type	Trichothecene	Mouse, Adult i.v.	Mouse, Adult i.p.	Mouse, Adult s.c.	Mouse, Adult p.o.	Mouse, Newborn s.c.	Rat i.v.	Rat i.p.	Rat s.c.	Rat p.o.	Guinea pig i.p.	Guinea pig s.c.	Guinea pig p.o.
A	T-2 toxin		5.2		10.5	0.15				5.2			3.06
	HT-2 toxin		9.0										
	8-Acetylneosolaniol							1.0					
	Neosolaniol		14 .5										
	NT-1		28.5										
	Diacetoxyscirpenol	12.0	23.0			0.17	1.3	0.73		7.3			
	Monoacetoxyscirpenol								0.725				
B	Nivalenol		4.1										
	Fusarenon-X	3.4	3.4	4.2	4.5	0.2				4.4	0.5	0.1	
	Diacetylnivalenol		9.6										
	Deoxynivalenol		70.0		46.0								
	3-Acetyldeoxynivalenol		49.0		34.0								

Type	Trichothecene	Rabbit i.v.	Cat s.c.	Dog i.v.	Dog s.c.	Swine i.v.	Duckling s.c.	Chick (p.o.) 1-day-old	Chick (p.o.) 7-day old	Trout p.o.
A	T-2 toxin		0.5			1.21		1.74	4.0	6.1
	HT-2 toxin							6.25		
	3′-Hydroxy-HT-2 toxin							8.5		
	Diacetoxyscirpenol	1.0		ca 1.1		0.37			5.0	
B	Fusarenon-X		5.0				ca 2.0			
	Deoxynivalenol				0.1		27.0			
	3-Acetyldeoxynivalenol				1.0					

Table 4
MINIMUM EFFECTIVE DOSE OF TRICHOTHECENES FOR INDUCING VOMITION[a]

Trichothecenes	Duckling[9,11]	Cat[5,8]	Dog[6,9]	Swine[10]	Human[12]
T-2 toxin	0.1	0.1			
HT-2 toxin	0.1				
Neosolaniol	0.1				
Diacetoxyscirpenol	0.2				2.4/*M*(i.v.)
Nivalenol	1.0				
Fusarenon-X	0.4	1.0	0.1(i.v.)		
Diacetylnivalenol	0.4				
Deoxynivalenol	13.5		0.1(i.v.)	0.05(i.p.)	
3-Acetyldeoxynivalenol	13.5		0.2		

Note: All values expressed in milligrams per kilogram. Dose is administered subcutaneously, except where indicated.

observed in isolated tissues such as fundus muscle, ileum, and atrium. This suggests that trichothecenes have no effects on the muscle cell membrane and the nerve elements of the tissues.

Diarrhea is one of the typical symptoms induced by the administration of a sublethal dose of trichothecenes. To clarify the mechanism of this symptom, pharmacological studies have been carried out using rats.[7] Intraperitoneal injection of fusarenone-X to rats caused watery diarrhea 36 to 60 hr after administration. Autopsy of animals 24 hr after the injection revealed an expansion of the small intestine, but no macroscopic bleeding into the lumen of the intestine was observed. In in vitro experiments, the toxin induced an increase in the absorption rate of D-xylose, but not D-glucose, from the intestine. Leakage of intravenously injected Evan's blue into the intestine was observed and the serum sodium ion level was decreased. However, leakage of erythrocytes into the intestinal lumen was not increased. Histological investigation revealed that the intestinal villi were shortened and caused an infiltration of the erythrocytes into the intestinal lamina propria. These findings suggest that the mycotoxin may increase the permeability of the blood vessel wall and intestinal mucosal epithelium, and cause leakage of plasma contents into the intestinal lumen.

Vomiting is one of the characteristic symptoms caused by trichothecenes. This symptom has been observed in cats,[5,8] dogs,[6,9] swine,[10] ducklings,[11] and sometimes, in human beings.[12] Table 4 summarizes the dose of trichothecene which induced vomiting in the animals. Vomiting was induced in dogs 5 to 15 min after i.v. injection of 0.3 mg/kg of fusarenon-X. When 0.9 mg/kg of the toxin was administered i.v., vomiting occurred 4.7 min after the injection and lasted for 2.5 hr. In ducklings, the minimum effective dose to induce vomiting by an s.c. injection of trichothecene ranged from 0.1 to 0.4 mg/kg, except for deoxynivalenol and its acetate. The effective dose of both deoxynivalenol and 3-acetyldeoxynivalenol was 13.5 mg/kg. When dog was used, however, the minimum effective dose of deoxynivalenol was the same as that of fusarenon-X. A pharmacological study[6] revealed that vomiting induced by 0.3 mg/kg of fusarenon-X in dog was antagonized by prior administration of 0.5 mg/kg of metaclopramide and 1.0 mg/kg of chloropromazine. These results indicated the possible action of trichothecenes on the chemoreceptor trigger zone in medulla oblongata.[6]

Chi et al.[13] investigated the level of biogenic amines after administration of T-2 toxin in chickens. Brain dopamine concentration was significantly increased and the brain norepinephrine was decreased in 4-week-old broiler chickens 4 to 48 hr after the intubation of 2.5 mg/kg of the toxin. The serotonin level in brain was not changed. Although the mechanism of change of biogenic amines in the brain is not clear, it is presumed that T-2 toxin caused

Table 5
MINIMUM EFFECTIVE DOSE OF TRICHOTHECENES FOR INDUCTION OF SKIN REACTION[16]

Trichothecenes	Mouse	Rat	Guinea pig	Rabbit
T-2 toxin	1		0.2	
HT-2 toxin	1		0.2	
Neosolaniol	10		1	
Diacetoxyscirpenol	10		0.2	1
Nivalenol	100		10	10
Fusarenon-X	10		1	10
Deoxynivalenol		75		
3-Acetyldeoxynivalenol		>1000		

Note: All values expressed in micrograms.

stressful stimuli and resulted in neural disturbances of the trichothecene-treated chicken characterized by abnormal positioning of wings, hysteroid seizure, and impaired righting reflex.

C. Dermal Toxicity

Skin necrotization is a characteristic biological activity of trichothecenes. Since dermal toxicity was introduced for screening toxic metabolites of *Fusarium* fungi, T-2 toxin and HT-2 toxin were isolated from cultre of *F. tricinctum.*[14,15] Minimum effective doses for the skin reaction are described in Table 5. T-2 toxin and HT-2 toxin are the most potent, while deoxynivalenol and 3-acetyldeoxynivalenol are less active in inducing skin reaction.

Histological investigations of cutaneous lesions induced by the application of low concentrations of T-2 toxin, diacetoxyscirpenol, and fusarenon-X revealed a nonspecific dermal inflammation characterized by hypermia, edema, and neutrophil exudation and necrosis of epidermis.[16,17]

A pharmacological study using fusarenon-X revealed that the vascular permeability of back skin of rabbit, estimated by exclusion of pontamin sky blue dye, increased biphasically with the maxima at 5 and 24 hr after application of the toxin.[18] This suggests that the increment of vascular permeability is one response of dermal tissues at early stage and chemical mediators such as histamine participate in the response to trichothecenes.

D. Hematological Changes

The first report on the hematological effect of trichothecene was that clotting and prothrombin times were increased after administration of crude T-2 toxin in rats.[19] The effects of trichothecenes on the number of blood cells were also studied using mice, rats, and cats.[8,20,21] The number of white blood cells was increased shortly after the administration of T-2 toxin in cats. In mice, similar effects were observed with T-2 toxin, neosolaniol, and fusarenon-X. Among leukocyte components, the greatest increase was observed in the number of lymphocytes followed by neutrophils. The number of circulating white blood cells was markedly decreased in cats when T-2 toxin was administered repeatedly: aleukia was induced by the toxin at a daily dose of 0.05 mg/kg (s.c.) for 1 week or 0.06 to 0.1 mg/kg (p.o.) for 24 days. The aleukia induced by T-2 toxin was also observed in chicks,[22-24] mice,[21] guinea pigs,[25] and monkeys.[26] In guinea pig administered with a daily dose of 0.9 mg/kg of T-2 toxin for 27 days, a decrease in the number of erythrocytes, a leukopenia, and an absolute lymphopenia were observed with a marked decrease in the number of lymphocytes in the bone marrow.[25] Diacetoxyscirpenol also induces aleukia in rats and dogs when administered i.v. at doses of 0.15 to 0.3 and 0.25 mg/kg, respectively.[27] These data indicated that T-2 toxin and other trichothecenes induce leukocytosis followed by advancing leukopenia in experimental animals. It is presumed that the trichothecenes have a high affinity for lymph nodes, spleen, and bone marrow, and short-term exposures of these tissues to the toxins induce rapid release of stored white blood cells; long-term exposures elicit disorder of the hematopoietic system of bone marrow and cause a gradual decrease of white blood cells.[28]

Recently, Segal et al.[29] reported that T-2 toxin induced hemolysis after a lag period and the mechanism of the hemolysis was suggested to be the same as a free radical mechanism.

E. Mutagenicity, Genotoxicity, and Teratogenicity

Recently, several short-term tests for screening carcinogenicity of chemicals have been developed. Rec-assay is one of the simplest and fastest tests. This assay is based on the fact that a mutant of *Bacillus subtilis* deficient in recombination is much more sensitive to agents attacking DNA than is the parent strain. Ueno and Kubota[30] examined the DNA-attacking abilities of T-2 toxin and fusarenon-X. Both toxins were negative in this assay system in the range of 20 to 100 μg per plate. *Salmonella* mutagenicity test (Ames' test) is also simple and fast. In this assay, no trichothecenes of *Fusarium* were positive in the presence and absence of 9000 × g supernatant fraction of rat liver.[31-33]

It is noteworthy that crotocin, a trichothecene of *Cephalosporium crotocigenum,* is mutagenic and enhances the mutagenicity of aflatoxin B_1 in this test system.[31] Crotocin has an additional epoxide group at C-7 and C-8 of the skeleton and this may be important for exerting the mutagenicity.

Eukaryotic cells are useful for evaluation of the genotoxicity of chemicals. Carcinogens induce DNA strand breakage as well as mutation in the cultured mammalian cells. In Hela cells and FM3A cells derived from C3H mouse mammary carcinoma, fusarenon-X, in the concentration of 1 to 32 μg/mℓ and 0.1 to 1 μg/mℓ, respectively, induced neither breakage of DNA strand nor the induction of 8-azaguanine-resistant mutation.[34] In bone marrow cells of Chinese hamster, T-2 toxin showed very weak clastogenicity; the bone marrow micronucleus test revealed a significant decrease of polychromatic erythrocytes.[35] Lafarge-Frayssinet et al.[36] demonstrated that T-2 toxin was capable of inducing single strand breakage of DNA in lymphoid cells in vivo and in vitro, and this breakage was not observed in hepatic tissue. These data suggest that the genotoxic effect of trichothecenes, estimated by breakage of DNA strand, depends on the cell line adopted.

Another study using an onion, *Allium cepa,* revealed that diacetoxyscirpenol induced chromosomal aberrations and cytogenic abnormalities in the root tips.[37]

Teratogenicity is another toxicological feature of trichothecenes. T-2 toxin administered intrapenetoreally to mice at doses of 1.0 and 1.5 mg/kg caused significant maternal mortality and decreased prenatal survivals. Fetal weight was decreased on days 7, 10, and 11. Gross malformation of the tail and limbs, exencephally with open eyes, and retarded jaw development were observed in the fetus when mother mice were treated on days 9, 10, or 11. Ribs and vertebral malformations were also detected in the fetuses exposed on days 9 and 10. Since trichothecenes are a potent inhibitor of protein synthesis in eukaryotes, such inhibition would cause a deterious influence on growth and survival of embryo. In addition, different distribution of the toxin in tissues, and differential sensitivity and cell cycle of particular organ primordial would cause the specific malformations.[38]

Another experiment was carried out using fusarenon-X.[39] Pregnant mice were administered with the toxin by s.c. injection or mixed with the diet. Abortion was induced by a single injection of 0.63 to 2.6 mg/kg with increasing frequency and shortening the latent period in proportion to increases in dose. In mice fed diets containing 5, 10, or 20 ppm of the toxin throughout the pregnancy or in the early phase of pregnancy, embryonal implantation was inhibited. Abortion was induced in all of the dams fed 20 ppm of the toxin for 7 days in the middle phase of pregnancy. In these experiments, no malformation was observed.

This contradictory observation between the experiments with T-2 toxin and fusarenon-X may be due to the difference in the chemicals, the strain of mice, and the route of administration.

Khera et al.[40] studied the embryonal toxicity of deoxynivalenol in mice. Dam mice were administered orally with the toxin on days 8 to 11 of pregnancy. A decrease in the rate of body weight gain in dams was observed in the 5, 10, and 15 mg/kg group; this was due to the resorption of embryo. In the 5 mg/kg group, the incidence of resorption was 80%, and the number of live fetuses and average fetal weight were reduced. Skeletal malformations such as lumbar vertebrate with fused arches or partly absent centra, absent or fused ribs, and missing, fused, or scrambled sternebrae were present in the 1, 2.5, and 5 mg/kg groups in a dose-related manner.

F. Chronic Toxicity

Marasas et al.[41] carried out a long-term feeding study using rainbow trout. T-2 toxin was fed in a diet at concentrations of 200 and 400 ppb for 12 months; any histological changes, including hepatoma, were observed. Noteworthy is the fact that the average weights and lengths of the toxin-fed trout were larger than those of the control group. Similarly, Ohtsubo and Saito[42] reported that no hepatomas or severe illness was observed in rats fed 10 ppm of T-2 toxin or fusarenon-X for 8 months. Saito and Ohtsubo[4] carried out a long-term feeding of T-2 toxin to mice. In mice fed T-2 toxin in a diet at 10 and 15 ppm, hyperkeratosis, acanthosis, and papillomatosis with inflammatory cell infiltration were observed in the forestomach after 13 weeks, but no squamous cell carcinoma was observed after 12 months.

Positive carcinogenic effects were reported by Schoental et al.[43] and Saito et al.,[44] although the incidence was very low. In the former experiment, Wistar-Porton rats were given T-2 toxin by a stomach tube at doses of 0.2 to 4 mg/kg alone or along with nicotinamide (200 to 250 mg/kg i.p.) at monthly or irregular intervals. Of 25 rats that survived 12.5 months, 19 had single or multiorgan tumors. They were brain tumor, pituitary adenoma and carcinoma, gastric papilloma and adenocarcinoma, pancreas adenoma, adenocarcinoma and islet cell adenoma, mammary carcinoma, and adrenal meduary tumor. In the latter, male Donryu rats were given a diet containing fusarenon-X at levels of 3.5 and 7 ppm for up to 2 years; 13 tumors, such as pituitary adenoma, gastric adenocarcinoma, thyroid adenocarcinoma, adrenocortical adenoma, pheochromocytomas, urinary papillary carcinoma, and leukemia, were found in 58 rats receiving the toxin, while 11 tumors were found in 45 control rats.

Trichothecenes exert dermal toxicity. Ueno and Shimada[45] examined the inducibility of skin tumor by T-2 toxin and fusarenon-X. On the back skin of mice, 0.1 to 1 μg of the toxins was painted twice a week for 1 year, but no papilloma was detected. On dimethylbenzanthracene-initiated mouse skin, these trichothecenes did not serve as promoters. Marasas et al.[41] also failed to produce papilloma by application of T-2 toxin with or without subsequent application of croton oil. On the other hand, Lindenfelser[46] concluded that T-2 toxin and diacetoxyscirpenol were not initiaters but weak promoters on dimethylbenzanthracene-initiated mouse skin after 23 weeks of application.

G. Cytotoxicity

Trichothecenes possess potent cytotoxicity to eukaryotes such as fungi, insects, animals, and plants. The specificity to eukaryotic cells is probably due to the high affinity for the eukaryotic ribosomes, especially to 60S subunit. The cytotoxicity of trichothecenes is closely related to the lethal toxicity to animals, cellular damages to actively dividing cells, impairment of immune response, and dermal toxicity in animals.

Most trichothecenes exhibit antifungal activity in a wide range of species.[47-49] Diacetoxyscirpenol was the first trichothecene from *Fusarium* reported to show antifungal activity. Spore germination of *Penicillium digitatum* was inhibited by 100 μg/mℓ of the toxin, but no inhibitory effect on *Bacillus subtilis, Salmonella dublin,* or *Escherichia coli* was observed.[3] T-2 toxin also inhibited the growth of 6 of 11 fungi, but had no effect on 54 bacterial strains at 4 μg per disk.[50]

As for phytotoxicity of trichothecenes, the first description was provided by Brian et al.[3] They isolated diacetoxyscirpenol from *F. equiseti* and *F. scirpi* as a phytotoxic substance during the screening of fungi for gibberellin production. Application of this trichothecene onto pea seedlings resulted in inhibition of stem elongation and scorched foliage. This toxin reduced growth of cress seedling roots and extension of pea internode section. T-2 toxin was also toxic to pea seedlings, and reduction in fresh weight and length were caused by immersion of the root in aqueous solutions of the toxin.[51] Phytotoxicity of T-2 toxin was demonstrated in growth of tobacco callus tissue,[52] germination of tobacco plant,[53] and auxin-promoted elongation of soybean hypocotyt.[54] Fusarenon-X was also demonstrated to inhibit the germination of *Brassia olearecea* seeds.[5]

A variety of non-mammalian animal species have been employed for toxicity studies of trichothecenes. A protozoa, *Tetrahymena pyriformis,* is sensitive to trichothecenes, and LC_{50} values of diacetoxyscirpenol, fusarenon-X, and nivalenol were 0.12, 5, and 10 μg/mℓ, respectively.[55] In a ciliate, *Colpidium campylum,* the minimum effective concentration of diacetoxyscirpenol was 0.5 μg/mℓ.[56] Trichothecenes were also highly toxic to brine shrimp *(Arthemia salina)* larvae,[57-61] fertilized eggs of sea urchin *(Hemicentrotus pulcherrimus),*[62] mature eggs of *Aedes aegypt;*[63] larvae of flurer beetle *(Tribolium confusum),*[64] a crustacean *(Cyclopus fuscus),*[65] zebra fish larvae,[66] a small fish Himedaka *(Oryzias latipes),*[5] chick embryo,[67] and larvae of amphibian *(Rana and Triturus).*[68]

In cultured mammalian cell systems, the cytotoxicity of trichothecenes has been demonstrated. Grove and Mortimer[69] studied the cytotoxicity of diacetoxyscirpenol and its chemically modified compounds to HEp2 of human origin and from baby hamster kidney (BHK). The lowest toxic concentrations of diacetoxyscirpenol were 5 and 1.5 ng/mℓ in HEp2 and BHK, respectively. This toxicity was eliminated by reductive opening of the 12,13-epoxy group and rearrangements of the trichothecene skeleton, indicating that the presence of the epoxide group and the complete ring system of trichothecene skeleton are important for the activity. Ohtsubo et al.[70,71] reported the cytotoxicity of nivalenol and related toxins to the HeLa cell. The cytotoxicity of T-2 toxin to epithelial cells of pig kidney, pk-15, was also demonstrated.[72]

To establish a rapid bioassay method for trichothecenes, Tanaka et al.[73] examined the cytotoxicity of 20 trichothecenes employing 3 lines of cultured cells: HeLa, HEK, and HL. The IC_{50} of trichothecenes from *Fusarium* species is shown in Table 6.

H. Immunological Disorder

Trichothecenes cause depletion of lymphoid tissue and this indicates that they induce disorder of the immune response. From this point, antibody response, cell-mediated immune responses (e.g., graft rejection, delayed hypersensitivity, and cell-mediated cytotoxicity), and mitogen-dependent proliferation of T cells have been studied.

The above immune responses are regulated by various types of cells in the lymphoid tissues. For example, antibody formation catalyzed by B-lymphocytes requires the help of T-lymphocytes, helper T-cell. For antigen recognition, the helper T-cells in turn are helped by antigen-presenting cells which are different from B- and T-lymphocytes. Therefore, if trichothecenes selectively deplete B-lymphocytes, the antibody production will be disturbed. If T-lymphocytes are depleted by the toxins, it is necessary to clarify what type of T-lymphocyte is depleted. The depletion of helper T-lymphocytes may reduce antibody formation, but that of suppressor T-lymphocytes may increase antibody production.

However, it is necessary to consider the fact that all the trichothecenes possess an inhibitory effect on protein and nucleic acid synthesis in actively dividing cells of eukaryotes. This property should cause modulation of immune responses, since antigen-stimulated lympho-

Table 6
CYTOTOXICITY OF TRICHOTHECENES TO CULTURED CELLS[73]

	IC$_{50}$ (μg/mℓ)		
Trichothecenes	**HeLa**	**HEK**	**HL**
Acetyl T-2 toxin	1.0	0.8	0.03
T-2 toxin	0.01	0.02	<0.003
HT-2 toxin	0.01	0.1	0.01
Neosolaniol	0.1	0.06	0.05
Diacetoxyscirpenol	0.01	0.01	<0.001
Monoacetoxyscirpenol	0.1	0.1	0.3
4,15-Diacetoxytrichothecene-3,7,8-triol	0.3	0.2	0.3
Calonectrin	3.0	5.6	0.3
Deacetylcalonectrin	7.0	10.0	1.0
Nivalenol	0.3	1.0	0.3
FusarenoneX	0.1	1.0	0.3
Deoxynivalenol	1.0	3.0	0.5
3-Acetyldeoxynivalenol	10.0	10.0	>10.0

cytes are capable of synthesizing protein and DNA to exert their function. Therefore, it is possible that the impairments of immune response are related to the inhibitory effect on macromolecule synthesis, and the potent immunological disorder ensured by trichothecenes cannot be simply discussed in parallel to potent immunosuppressors such as cyclophosphamide and cyclosporin A, as reviewed by Mirocha[74] and Ueno.[18]

1. Antibody Formation

Richard et al.[24] reported that dietary T-2 toxin caused atrophy of the thymus in turkey, and Sato et al.[21] observed a marked decrease in the weight of the spleen and thymus in mice after intubation of 3 mg/kg of T-2 toxin for 3 and 5 days. Rosenstein et al.[75] assessed antibody responses by counting antibody-forming cells in the spleen of immunized mice. Mice were fed 0.7 mg/kg of T-2 toxin for 7 days, and sheep erythrocytes were injected on day 3 after the first treatment of the toxin. On day 8, a marked reduction of antibody-producing cells in the spleen as well as complete suppression of serum hemagglutinin titers were observed. This immunosuppressive effect of T-2 toxin disappeared 6 days after termination of the toxin.

Masuko et al.[76] and Otokawa et al.,[77] however, found that a single dose of 3 mg/kg of T-2 toxin in mice caused modification of delayed hypersensitivity responses without affecting antibody response. This apparent contradiction is explained by a different timing in administration of T-2 toxin, in which mice received the toxin once within several days before or after antigen stimulation. Experiments using fusarenon-X revealed that both IgE and IgG1 antibody response to dinitrophenyl ovalbumin were suppressive when mice were repeatedly treated with the toxin at doses larger than 25 μg/day before antigen stimulation.[78] In mice stimulated with pokeweed mitogen and lipopolysaccharide (LPS), in vitro antibody production in spleen cells from mice treated with fusarenon-X was also suppressed.

Sato et al.[79] examined the effect of fusarenon-X on antibody responses to vaccines of Newcastle disease in chicks. After hatching, the toxin was fed at 8 ppm in feed for 6 weeks, and live or inactivated vaccine was inoculated at 2 weeks of age. Hemagglutinin inhibitory antibody titers were reduced by the toxin treatment in chicks immunized with the live vaccine.

2. Allograft Rejection

There have been several evidences of inhibition of cellular immunity by trichothecenes. Rosenstein et al.[75] examined the effect of T-2 toxin on the mean survival time of the grafted skin from C57BL/6 mice to Swiss mice. The mean survival time of the graft was 12 days when the recipients were treated with 0.75 mg/kg/day of the toxin for 7 days before the skin graft and then 3 times a week for 20 days, while that of control mice was 8.69. This means that T-2 toxin suppressed certain steps of immunity in allograft rejection.

3. Delayed Hypersensitivity

Delayed hypersensitivity (DH) is an immune response mediated by sensitized T-lymphocytes. Otokawa et al.[77] studied the effect of T-2 toxin on DH response in mice to clarify the possible impairment of T-lymphocytes. Mice were sensitized by s.c. injection of sheep erythrocytes (SRBC) and the DH response was estimated by foot pad swelling 7 days later. When mice received 3 mg/kg of the toxin before or on the same day as sensitization, no appreciable effect on DH was observed. However, when the toxin was administered 2 or 3 days after the antigen sensitization, a marked enhancement of DH response was observed. This indicates that the timing of toxin exposure is critical for the enhancement of the DH response. The optimal time for toxin administration was 2 days after the sensitization. Since the life time of effective concentration of T-2 toxin in vivo is very short, and the optimal timing of toxin administration corresponds with time of appearance of suppressor cells, it is presumed that trichothecenes may interfere with the proliferation of suppressor T-cells for DH.

4. Blastogenic Response to Lectins

Certain mitogens, such as phytohemaglutinin (PHA) and concanavalin A (con A), stimulate the proliferation of T-cells in vitro, and LPS induces the same phenomena in B-cells. Lafarge-Frassinet et al.[80] investigated PHA and LPS responses of lymphocytes from mice treated with trichothecenes. Animals were treated with crude T-2 toxin and diacetoxyscirpenol and the mitogen responsiveness of spleen and thymus cells was examined. The data indicate that stimulation of both T- and B-cells was inhibited reversively, and the ability to synthesize anti-SRBC antibodies was suppressed. In vitro effects on lymphocytes and fibrosarcoma cell cultures caused a direct cytostatic action at high concentrations and a stimulating action at low concentrations. Histopathological observations indicated severe damage only in the thymus and spleen. These results suggest that the immune system is sufficiently sensitive to trichothecenes to be impaired at doses not inhibitory for other organs. Decreased resistance to infection may occur as a result of immunosuppression.

5. Infection

It is highly possible that the immunosuppression caused by trichothecenes results in an increased infection in animals. Boonchuvit et al.[81] studied the effect of T-2 toxin on *Salmonella* infection. Chickens received a diet containing 16 ppm of the toxin for 1 week and were inoculated orally with 1×10^8 cells of *Salmonella.* T-2 toxin caused an increased mortality in the chicken. Similarly, Kanai and Kondou[82] observed that T-2 toxin enhanced the mortality of mice infected with *Mycobacterium obvis.* Available information on the relationship between trichothecenes and the immune response is not sufficient at present, and it is difficult to generalize the toxic effect of trichothecenes on this biologically important defense system. Since various types of lymphocytes participate sequentially in the immune responses, the impairments depend on the doses of toxins, the timing of administration, and the sensitivity of respective cells.

I. Biochemical Activity

1. Inhibition of Protein Synthesis

The initial observation that trichothecenes inhibit protein synthesis in mammalian cells was made in the author's laboratory.[83] Nivalenol was shown to induce a dose-dependent inhibition of incorporation of several amino acids into protein in rabbit reticulocytes[83] and Ehrlich ascites tumor cells.[84] This inhibitory effect of trichothecenes was demonstrated in whole animal,[85] a protozoa,[55] HeLa cells,[86] cultured mammalian cells,[70,71] rat liver preparations,[87] hamster ovary cells,[88] human tonsil,[89] yeast spheroplasts,[90,91] rabbit reticulocytes,[2,83,87,93-97] and lymphocytes,[98] but not in bacterial cells.[87] There have been active investigations to elucidate the molecular mechanism of inhibition of protein synthesis by trichothecenes. Since nivalenol was found to inhibit poly U-directed incorporation of phenylalanyl-tRNA without affecting the activation of amino acids, the inhibition was presumed to be caused by impairments of ribosomal function.[84]

Other approaches were made through analyses of polyribosomal profile in rabbit reticulocytes and cultured cells. Fusarenon-X caused the breakdown of polysomes shortly after its exposure to rabbit reticulocytes[87] and mouse fibroblasts (L-cells).[99] Breakdown of polysomes is associated with the inhibition of initiation step of protein synthesis. T-2 toxin and diacetoxyscirpenol also induced the disaggregation of polysomes in Hela cells, while no breakdown was observed with trichodermin, a trichothecene of *Trichoderma viride*.[86] This result was consistent with the effect of an elongation or termination inhibitor of protein synthesis.[92] From these analyses of polysomal profile, the trichothecenes were classified into two types: one is an initiation inhibitor (I-type) and the other is an elongation-termination inhibitor (ET-type).[86,91,92,95,100] The former includes nivalenol, fusarenon-X, monoacetoxyscirpenol, T-2 toxin, HT-2 toxin, diacetoxyscirpenol, and calonectrin, and the latter includes deoxynivalenol, 15-deacetylcalonectrin, and trichodermin.

This classification of trichothecenes is somewhat arbitrary because experimental conditions, such as concentration of toxins in an assay system, have a great influence on disaggregation or stability of the polysome. The important point is that trichothecenes, regardless of whether they are of I-type of ET-type, possess an ability to interact with the peptidyl transferase center on the 60S ribosomal subunit and inhibit the transpeptidation or peptide bond formation process. Several investigations revealed that neither trichodermin (ET-type) nor T-2 toxin (I-type) affected the ribosomal binding of release factor, elongation factor-2, and aminoacyl-tRNA or inhibited the ribosome-dependent GTPase activity of elongation factor-2 and release factor in reticulocyte lysate.[101] Further experiments indicated that T-2 toxin (I-type) did not inhibit the assembly of the 80S initiation complex, but prevented the release of methionin by puromycin.[102] Wei et al.[103] and Cannon et al.[104] demonstrated that trichothecenes possess affinity for ribosomes and their 60S subunit. Good correlation was observed between the inhibition of termination and peptidyl transferase activity by trichodermin, and that of trichodermin binding by other trichothecenes. One trichothecene binding site per ribosome corresponds to one transferase catalytic center.[92,104,105]

Recent approaches have been focused on clarifying the molecular species of the ribosomal protein which regulates the binding of trichothecene. Mutants of yeast[105-109] and Chinese hamster cells[88] have been isolated which were resistant to the effects of trichothecenes on protein synthesis in whole-cell and cell-free system. The binding of labeled trichodermin to mutant cells, ribosomes, and 60S subunits was reduced, indicating that 60S subunits are associated with the resistance to the toxin. In the resistant yeast cell, a single recessive nuclear gene was responsible for resistance, and this resistance locus, tcm^{-1}, was found to be closely linked to the locus *pet* 17 on the right arm of chromosome XV.[106] Yeast cells sensitive to trichodermin have been transformed with autonomously replicating recombinant plasmids carrying DNA fragment of the genome of a trichodermin-resistant yeast strain. After selection for trichodermin-resistant cell, several transformants yielded a plasmid containing a 13.5-kilobase DNA fragment that encoded the resistance gene tcm^{-1}, and the gene for ribosomal protein L3.[110] This result clearly demonstrated that the gene for trichodermin resistance in yeast specified ribosomal protein L3, the largest of the yeast ribosomal proteins. An interesting finding is that ribosomal subunits of *Myrothecium verrucaria,* one of the producers of macrocyclic trichothecenes, were suspected to resist T-2 toxin. This also means that 60S subunits of eukaryotes are repsonsible for the sensitivity to trichothecenes, and the ribosomes of trichothecene-producing fungi are genetically resistant to the toxins they produce.[111]

As mentioned above, trichothecenes are one of the potent inhibitors of protein synthesis in eukaryotic cells, and this characteristic property of trichothecene is expected to be a valuable tool for the analyses of biological reactions occurring in eukaryotes. For example, using T-2 toxin as an inhibitor of protein synthesis, Sun et al.[112] analyzed cell-free synthesis of T1 globulin (legumen) in the polysomes prepared from developing cotyledons of French bean (*Phaseolus vulgaris* L.), and elucidated that 29% of the total synthesis depended on initiation events.

2. Inhibition of Nucleic Acid Synthesis

Dose-dependent inhibition of DNA and RNA synthesis by nivalenol was first observed with Ehrlich ascites tumor cells.[95] Apart from the marked (70%) inhibition of protein synthesis with 1 to 10 μg/mℓ of nivalenol, thymidine incorporation into DNA was inhibitied up to 60%, but a relatively weak suppression (30%) of uracil incorporation into RNA was observed. A similar result was demonstrated in cultured cells such as KB cells,[70] HeLa cells,[86] mouse L-cell fibroblast,[99] hamster ovary cells,[88] and lymphocytes[98] with trichodermin, diacetoxyscirpenol, fusarenon-X, and others. From the evidences that the inhibition of DNA and RNA synthesis required a higher concentration of toxins and the extent of inhibition was much less as compared with that in protein synthesis, the observed inhibition of DNA and RNA synthesis in toxin-treated animal cells was presumed to be a secondary effect of trichothecenes. Tashiro et al.[113] found that fusarenon-X in concentrations of 0.0236 to 1.889 μmol/mℓ was neither inhibitory to DNA-dependent RNA polymerase of rat liver nuclei and *E. coli* nor suppressive to nuclear RNA hybridase of rat liver and *Tetrahymena pyriformis* in vitro, and these data supported the hypothesis mentioned above.

As for the detailed mechanism of impairment of nucleic acid synthesis, no final conclusion has been presented. According to Munsch and Mueller,[114] inhibition by T-2 toxin of incorporation of thymidin into DNA in a cell line derived from thymus was strong in concentrations over 1 ng/mℓ and weak in 0.1 to 10 ng/mℓ. The toxin at low concentration (0.1 to 1 ng/mℓ) caused a transient increase of DNA polymerase alpha- and beta-terminal deoxynucleotidyl transferase, while the toxin strongly inhibited these enzyme activities at a higher concentration (over 1 ng/mℓ).

Agrelo and Schoental[115] reported that hydroxyurea did not affect unscheduled DNA synthesis in the cells treated with T-2 toxin and HT-2 toxin, but the combination of rat liver microsomes and hydroxyurea resulted in an increase in the unscheduled DNA synthesis in the cells exposed to 100 μg/mℓ of HT-2 toxin. This was often observed with potent carcinogens such as benzo(*a*)pyrene, which requires cytochrome P-450-dependent metabolic activation to modify DNA and gene expression. At present, the detailed mechanism of action and metabolism of trichothecenes has not been elucidated, but the data of Agrelo and Schoental represents a possibility of participation of the microsomal drug metabolizing enzyme in the induction of DNA damage by trichothecenes.

A protozoa, *Tetrahymena pyriformis,* is a useful model for analyses of eukaryotic cells. In this organism, fusarenon-X and related toxins inhibited protein and nucleic acid synthesis in a manner similar to that observed in cultured mammalian cells.[84] Recent observations with synchronously dividing *Tetrahymena* cells revealed that the incorporation of labeled amino acid, thymidine, and uracil into protein, DNA, and RNA, respectively, was affected by nearly the same concentration of T-2 toxin.[115] The molecular mechanism of inhibition of protein synthesis was clearly explained by the high affinity of labeled T-2 toxin for 60S ribosomal subunits of this protozoa as is in cultured cells. However, neither DNA and RNA synthesis nor RNA hybridase activity was affected by T-2 toxin in the in vitro system using isolated nuclei from normal cells or from cells pretreated with the toxin. The mechanism of inhibition of nucleic acid synthesis in the in vitro system is not clearly understood. An experiment using cytochalasin B, a potent inhibitor of cellular membrane functions, revealed that the effects of this mycotoxin were similar to those observed with trichothecenes. Therefore, it is probable that some membrane components which regulate nuclear RNA synthesis may be affected by trichothecenes.

3. *Other Biochemical Effects*

Chiba et al.[117] found that fusarenon-X inhibited the uptake of phosphate in *Tetrahymena* cells. This inhibition may suggest that trichothecenes affect membrane functions. The toxin also caused 50% inhibition of incorporation of acetate into phopholipids and tenfold stimulation of incorporation into triglyceride. These effects may be secondary to the inhibition of phosphate uptake. In rabbit reticulocytes, however, fusarenon-X, T-2 toxin, and neosolaniool in a concentration of 100 μg/mℓ were not inhibitory to Na ion-dependent glycine transport.[118]

In general, epoxide groups spontaneously or enzymatically react with thiol groups. This reactivity of the epoxide groups may be related to the toxicity of trichothecenes. When SH enzymes were pre-incubated with selected trichothecenes in the absence of their substrate, the activities of such enzyme as creatine phosphokinase, lactate, and alcohol dehydrogenases were decreased.[119] However, neither urease activity[120] nor rat liver lysosomal cathepsin activity[121] was affected by diacetoxyscirpenol or T-2 toxin, respectively, in vitro. Foster et al.[122] reported the reactivity of T-2 toxin with glutathione in the presence of epoxide-glutathione-*S*-transferase.

When whole animals such as mice were exposed to trichothecenes, no detectable alteration was observed in liver and kidney functions. In chicks fed a diet containing 10 ppm of T-2 toxin for 3 weeks, a hepatic microsomal aminopyrine demethylase was reduced, whereas another mixed function oxydase, aniline hydroxylase, was not significantly affected.[142] In mice administered with a sublethal dose of fusarenon-X, a rapid hypoglycemia was followed by depletion of hepatic glycogen.[124] It was suggested that the toxin induced a malfunction of glucose absorption in the intestine and caused an accelerated glycolysis. But as mentioned above, trichothecenes are highly cytotoxic to mucous membrane epithelia of the intestine, and this cytotoxicity may result secondarily in these impairments of carbohydrate metabolism and absorption.

III. ZEARALENONES

An estrogen produced mainly by *F. graninearum* (*Gibberella zeae* in perfect stage) induces a marked uterotropic response in livestock and experimental animals. Estrogenism in swine was first reported by McNutt et al.[125] in 1928. Sows which had consumed moldy feed developed enlarged vulvae and rectum. McErlean[126] reported an association of barley infected with *Fusarium* fungi with an estrogenism in swine in Ireland. Dorfman and Dorfman[127] observed abnormal weights of uteri, which were caused by estrogen present in a ration supplemented with molded field corn. Stob et al.[128] isolated an anabolic and uterotropic compound from corn infected with *Fusarium* which appears to be responsible for the estrogenic syndrome. Christensen et al.[129] isolated an estrogenic metabolite, F-2, from a culture of *Fusarium,* and Urry et al.[130] elucidated the structure of F-2 as 6-(10-hydroxy-6-oxo-*trans*-1-undecenyl)-β-resorcyclic acid-μ-lactone, and named it "zearalenone". Kuo et al.[131] and Taub et al.[132] determined the absolute configuration around the optically active asymmetric carbon, C′-10, of the natural enantiomorph as S (senister). Total synthesis of zearalenone was performed by Taub et al.,[132] Girotra and Wendler,[133,134] Vlattas et al.,[135] Hurd and Shah,[136] and Masamune et al.[137]

Mirocha et al.[138] reported eight naturally occurring derivatives of zearalenone. They were isolated from *F. roseum* cultured on corn and designated as F-5-0 through F-5-7. F-5-3 and

F-5-4 were elucidated to be 8′-hydroxyzearalenone and 8′-epihydroxyzearalnone, respectively,[139] and F-5-1 and F-5-4 were epimeric mixtures of 3′-hydroxyzearalenone.[138] F-5-0 was 4′,5′-dihydroxyzearalenone. 6′,8′-Dihydroxyzearalenone[140] and α-zearalenol[141] were isolated from a culture of *F. roseum*. Bolliger and Tamm[142] isolated 5-formylzearalenon and 7′-dehydrozearalenone along with epimers of 8′-hydroxyzearalenone from *Gibberalla zeae*. Recently, Ellastad et al.[143] isolated zearalenone-like macrolides from an unidentified fungus, and one of them was elucidated to be (4′S, 5′S)-4,5′-dihydroxyzearalenone 4-methyl ester. The chemical structures of zearalenone and its derivatives are shown in Table 7.

A. Toxicity

The acute toxicity of zearalenone obtained from single-dose experiment, expressed as LD_{50} (mg/kg i.p.), was 500 (female mice), 5490 (male rats), and 2500 (female guinea pigs). The oral LD_{50} of zeralenone was over 5 to 20 g/kg in mice, rats, and guinea pigs.[144] This means that the oral administration of a single dose of 2 g/kg does not cause death in mice and rats. Maximum dose induced slight irritability and hypoactivity during the observation period of 14 days. Intraperitoneal administration of the compound induced irritability, bradynea, hypoactivity, and hypothermia. In monkeys and dogs, no significant general drug action was observed in behavioral changes, food-deprived tests, electrolyte excretion, GI motility, blood pressure, and neurophysiological activities.

Bailey et al.[145] carried out acute and subacute toxicity test with male and female rats. The acute single-dose lethality could not be determined in the rats but was higher than 16 g/kg. In the subacute toxicity test, rats were given 0, 0.1, 1, and 10 mg/kg/day. Growth and food intake were significantly depressed for both males and females in the 10 mg/kg group. Fertility was greatly impaired in females given a high dose. Only 26 pregnancies occurred out of the 50 matings in the F1A litter and only 9 out of 22 in the F2A litter. The number of resorptions and stillborns was greatly increased with 56% of the high-dose dams showing complete litter resorption. Fetal examination indicated no teratogenic responses in any level of zearalenone. These data indicated that zearalenone decreased fertility and reproductive ability, increased the number of resorptions and stillborn fetuses, and increased medullary trabecula in those born. The noneffective level was 0.1 mg/kg.

Ruddick et al.[146] reported teratogenicity of zearalenone in rats. Female Wister rats were mated and were treated on days 6 through to 15 of gestation by intubation with 0, 1, 5, and 10 mg/kg/day of zearalenone. On day 22 of gestation, all females were anatomized. None of the dosages produced any apparent signs of maternal toxicity during the gestation period. The incidence of delayed or absent skeleton ossification increased with increasing dosage; the rib, sternum, tarsal bones, and the partietal bone were commonly affected. At a dose of 1 mg/kg, the incidence of skeletal defects was 12.8%, and it was increased to 26.1 and 36.8% at 5 and 10 mg/kg, respectively. These incidences were much higher than generally observed (5%) for this strain of rats.

Ruzas et al.[147] investigated the effect of zearalenone on sexual differentiation in rats. The offspring of rats fed with *Fusarium graminearum*-infected corn during pregnancy showed a significant decrease in fertility in both sexes. This reduced fertility was even more pronounced when newborn rats were injected with zearlenone at a dose of 30 μg per animal. These data suggested that zearalenone might affect sexual differentiation with a mechanism similar to androgen or estrogen sterilization.

In addition to the estrogenic activity, the cytotoxic nature of zearalenone was reported. Vanyi et al.[148] examined cytotoxic effect on various monolayer cell cultures. The testicular cell cultures prepared from animals whose spermatogenesis was inhibited by zearalenone proved to be more sensitive than those from resistant species. Horakova et al.[149] studied the cytotoxicity on cultured HeLa cells and the IC_{50} was 7.5×10^{-5} *M*. Vianello et al.[150] investigated the effect of zearalenone on plant tissues. The compound was found to stimulate a leakage of electrolytes and inhibit rubidium uptake in roots of *Zea mays* and *Beta vulgaris*

Table 7
NATURALLY OCCURRING DERIVATIVES OF ZEARALENONE

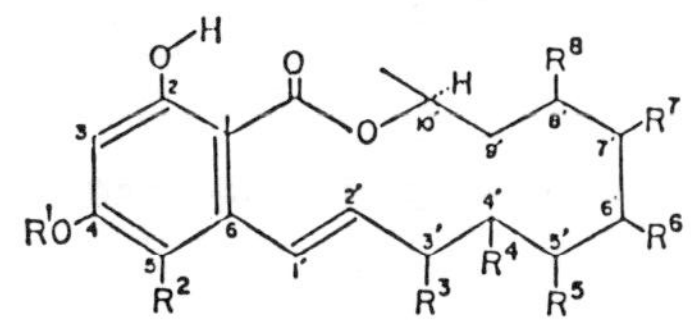

Name	R^1	R^2	R^3	R^4	R^5	R^6	R^7	R^8
Zearalenone	H	H	H_2	H_2	H_2	=0	H_2	H_2
α-Zearalenol	H	H	H_2	H_2	H_2	OH(R)	H_2	H_2
8′-Hydroxyzearalenone (F-5-3)	H	H	H_2	H_2	H_2	=0	H_2	OH(R)
8′-*epi*-Hydroxyzearalenone (F-5-4)	H	H	H_2	H_2	H_2	=0	H_2	OH(S)
3′-Hydroxyzearalenone (F-5-1)	H	H	OH	H_2	H_2	=0	H_2	H_2
3′-Hydroxyzearalenone (F-5-2)	H	H	OH	H_2	H_2	=0	H_2	H_2
6′,8′-Dihydroxyzearalene	H	H	H_2	H_2	H_2	=0	H_2	OH
4′,5′-Dihydroxyzearalene (F-5-0)	H	H	H_2	OH	OH	H_2	H_2	H_2
5-Formylzearalenone	H	CHO	H_2	H_2	H_2	=0	H_2	H_2
7′-Dehydrozearalenone	H	H	H_2	H_2	H_2	=0	H	H
LL-Z1640-1	CH_3	H	H_2	OH(S)	OH(S)	=0	H_2	H_2
LL-Z1640-2	CH_3	H	H_2	OH	OH	=0	H	H
LL-Z1640-3	CH_3	H	H_2	OH	OH	OH	H_2	H_2
LL-Z1640-4	CH_3	H	H_2	OH	OH	OH	H	H

var. *rubra*. Furthermore, the toxin caused inhibition of hydrogen ion extrusion, elongation, ATPase activity in plasmolemma-enriched fraction, and depolarization of transmembrane potential of corn root. These data indicated an impairment of plasma membrane of several plant species. As for antibacterial activity, zearalenone showed a limited activity to *Bacillus thuringiensis* in a concentration of 2500 μg/mℓ.[151]

Recently, carcinogenicity testing of zearalenone was carried out by the National Cancer Institute (NCI).[152] The test was conducted by feeding diets containing 25 and 50 ppm of zearalenone to groups of 50 F344 rats of each sex and 50 or 100 ppm to groups of 50 B6C3F1 mice of each sex for 103 weeks. Control groups consisted of 50 rats and 50 mice of each sex. Calculation of feed consumption revealed that the rats of the low- and high-dose groups received about 1 and 2 mg/kg/day of zealarenone, respectively, during most of the research. The mice of the low- and high-dose groups received 7 to 10 and 14 to 20 mg/kg/day, respectively.

Histopathological examination revealed that in rats, inflamation of the prostate, testicular atrophy, hepatocellular cytoplasmic vacuolation (in male), and nephrosis in both sexes were dose dependent. Pituitary adenomas were observed in 3 of 10 females given 50 ppm of zearalenone in the 13-week subchronic study, but tumor development was not related to the dose of the chemical. In mice, myelofibrosis in bone marrow, uterine fibrosis, and cystic ducts in mammary glands were dose dependent in the females. An interesting finding was the dose-dependent incidence of hepatocellular adenomas in the females.

Hepatocellular adenomas are considered to be a part of the process of hepatic carcinogenesis, and a chemical that induces adenomas represents a potential carcinogenic risk to humans. Despite the significant similarity in the physiological properties of zearalenone and diethylstilbestrol (DES), there appears to be significant differences in tumor target tissues. In Sprague-Dawley rats orally administered with DES, pituitary tumors, hepatomas, and a hemangio-endothelioma were evident.[153] These tumorous changes were not observed in the F344 rats. In mice given DES orally, mammary tumors, uterine horn adenocarcinomas, cervical adenocarcinomas, and esteosarcomas developed.[154] The B6C3F1 mice used in this experiment did not develop malignant changes in these tissues. It is still a matter of doubt whether the difference of their target tissues is due to strain or other experimental conditions. The NCI report concluded that the administration of zearalenone was associated with increased incidences of hepatocellular adenomas in female B6C3F1 mice and of pituitary adenomas in male and female B6C3F1 mice in a dose-dependent manner; thus, zearalenone should be considered as a carcinogen in male and female mice.[152]

Zearalenone and α-zearalenol induced a positive effect in rec-assay, while β-zearalenol was negative. Further experiments with the *Salmonella* mutagenicity test indicated that these zearalenone derivatives were not mutagenic in the presence and absence of 9000 × g supernatant fraction of rat livers.[34] Negative mutagenicity of zearalenone and its derivatives was also reported by other groups.[32,155] Therefore, if the conclusion of the NCI is correct, no positive correlation between the carcinogenicity and mutagenicity was demonstrated as far as zealarenone is concerned. Similar negative effects were reported for other anabolic compounds. Although anabolic steroids have proved to be carcinogens in several animals species,[156] they are not mutagenic in the bacterial test system.[157] Recent data revealed that zearalenone and other anabolic compounds, such as 17β-estradiol, possess a high affinity for the steroidal receptor protein of the cytoplasmic fraction of the target tissues. The present concept of carcinogenicity of sex hormone strongly suggests an involvement of the cytoplasmic receptor protein, which is not present in the bacterial testing system.

B. Mechanism of Action

Zearalenone and its derivatives evoke estrogenic effect in a variety of assay systems using mouse, rat, rabbit, chicken, ewe, pig, and monkey, as reviewed by Hidy et al.[145] In immature female mice, zearalenone administered orally in doses of 1 to 10 mg/kg induced a marked increment of uterine weight, and in ovariectomized female adult mice, 2 mg/kg/day for 1 week induced a significant increase in uterine weight.[158] In young female pigs, daily intubation of 1 mg/kg of zearalenone causes a marked enlargement of vulvae in 5 days.[159] Hobson et al.[160] studied the estrogenic effect of zearalenone in the rhesus monkey, which was used as a model of non-human primates. The injection of zearalenone initially caused a decrease in the gonadotropin level, followed by that in lutenizing hormone (LH), but not always in follicle-stimulating hormone (FSH) surge. This pattern was similar to the effect of DES and 17β-estradiol. In ovariectomzied monkeys, the minimum doses of zearalenon, DES, and 17β-estradiol which depressed LH levels after s.c. injection were 14, 0.5, and 4 μg/kg, respectively, and those after oral administration were 56, 2, and 1 μg/kg, respectively. These findings indicated that the estrogenic activities of these three compounds were quite similar to each other when given s.c., but the physiological activity of zearalenone was much lower than that of the other estrogens when administered orally. This may suggest a low absorption of zearalenone when given orally. Additional experiments for clarifying whether DES or zearalenone would interfere with either the estradiol-induced gonadotropin depression or the subsequent gonadotropin peak revealed that progesterone blocked the induction of LH peaks by estradiol injection, but neither zearalenone nor DES affected the LH peak or the depression.

17β-Estradiol and other estrogenic drugs exert their physiological activity firstly by binding with cytosol receptor protein and secondly by translocation of the hormone-receptor complex to nuclei. Interaction between zearalenone and estrogen receptor has been reported from several laboratories.[161-169] Most of the experiments were carried out using the estrogen receptor of rat uterus,[162,164,168,169] though some experiments were performed with mouse uterus,[162,167] calf uterus,[161,164] rat mammary gland,[162] human breast cancer cells,[165] rat brain,[170] human mammary tumor tissues,[169] and 7,12-benz(*a*)anthracene-induced rat mammary tumors.[168] Competition experiments revealed the affintity (K_d) of zearalenone and its derivatives to the cytoplasmic estrogen receptor of rat uterus as follows: α-zearalenonl, 3×10^{-9}; zearalenone, 7.7×10^{-8}; β-zearalenonl, 5×10^{-7}; 8′-hydroxyzearalenone, 13.5×10^{-6} *M*.[186] These data indicate that the estrogenic activity, estimated by the affinify for the estrogen receptor, is influenced by the 6′-hydroxyl group, and α-epimer is much more active than the β-epimer. This biochemical feature of zearalenone and its derivatives reflects their in vivo estrogenic activities estimated by the increment of uterine weight and other parameters.

In the early stage of investigation on the molecular basis of zearalenone action, biochemical modulation of uterine tissues of mice exposed to zearalenone was analyzed in the author's laboratory.[171] In the isolated uteri of mice administered with the compounds, a marked increase in the uterine permeability for uridine and amino acids was observed, and this biochemical change was followed by an increment of protein and nucleic acid synthesis. These changes were quite similar to the effect of natural estrogen, estradiol. Further accumulated data revealed that zearalenone possesses a high affinity for cytoplasmic estrogen receptor, and this receptor-binding zearalenone translocates to nuclear site to modify the expression of genetic information. The early event is the induction of specific protein (IP);[172] this IP induction takes place in the uterus of animals administered with zearalenone and also in the same tissue incubated with the compound in in vitro systems. The molecular weight of this zearalenone-induced protein was similar to that induced by estradiol. Therefore, it is deduced that zearalenone binds with the uterine cytoplasmic receptor protein; this zearalenone-receptor complex is translocated to the nuclei to lead the induction of IP and thereafter result in increase in activities of nuclear RNA polymerase I and II and of protein synthesis. Thus, the mode of action of zearalenone is considered to be mediated through its association with estrogen receptor in the target tissue, as in the case of estradiol, although it is not clear whether zearalenone itself or its active metabolite occupies the binding site of nuclei.

Recently, the interaction between zearalenone and brain estrogen receptor in rats was investigated in the author's laboratory.[170] The data showed that zearalenone is transferred

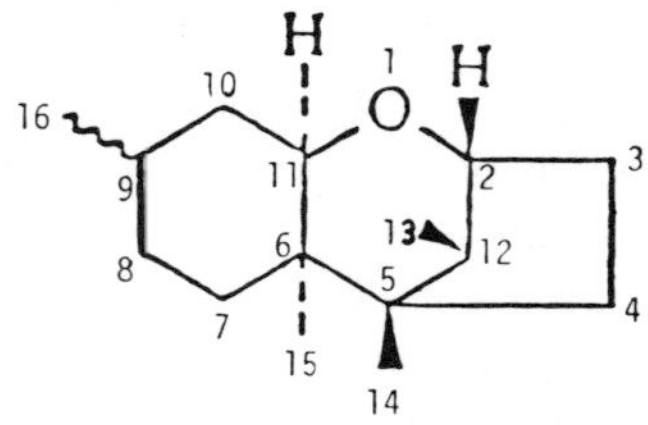

FIGURE 1. Trichothecane.

to the rat brain through blood barrier and binds with the estrogen receptor of hypothalamus and hypophysis. This result suggests that zearalenone affects the estrogen feedback system through the estrogen receptor of the brain in rats.

As for the metabolism of zearalenone, two metabolites (I, II) in the feces and urine, respectively, and two gluculonides in the urine were detected as the metabolites in rats intubated with the compounds.[173] Metabolite I was identified as α-zearalenol, which was one of the biotransformed products of zearalenone by the hepatic enzyme.[174] In pigs, both free and conjugated zearalenone (63%), α-zearalenol (32%), and β-zearalenol (5%) were detected in the urine.[175] Biochemical analysis of the metabolic pathway of zearalenone revealed that NADH- and NADPH-dependent zearalenone reductase(s) are localized both in the cytosol and the microsomes of rat liver, and more than 90% of the total activity is tightly bound with the microsomes. This microsomal reductase possesses a relatively high molecular weight and is resistant to SH-reagents and heavy metal ions.[176] These enzymological properties of the microsomal zearalenone reductase(s) are largely different from those of 3-hydroxy steroid dehydrogenase and other keton reductases. In addition, there are species differences in the zearalenone reductase activity in regard to the stereospecificity of the reduction. This multiplicity of zearalenone reductase may explain why the relative ratio of α- and β-zearalenol production are different among livestock.[177]

IV. BUTENOLIDE

Butenolide is reported as one of causes of tall fescue forage toxicity problems. Livestock such as cattle grazing on fescue pasture are sometimes affected with "fescue foot", a noninfectious disease characterized by lameness of the hind quarters and a predisposition to dry gangrene of the extremities. Although the symptoms are similar to those of ergotism, ergot is not involved in fescue foot, as reviewed by Yates.[178]

Tall fescue toxicity is associated with the unknown chemical(s) in the toxic forage. In 1959, scientist at the Northern Regional Research Laboratory prepared an alcohol extract of some toxic fescue hay that was furnished by the Kentucky Agricultural Experimental Station. This extract induced typical symptoms of fescue foot in a fisturated heifer. This was the first time that an extract of fescue hay was shown to be toxic to cattle. By employing a novel bioassay with tail temperature change, an unknown vasoconstrictor was postulated as the causative agent of this type of foodborne poisoning.[179]

After the finding that alimentary toxic aleukia (ATA) was caused by mycotoxins in cereal grains and these toxins possess a high dermal toxicity, Keyl et al.[180] applied this dermal response in rabbits as a screening bioassay. Of 24 isolates from a toxic hay, 9 were proved to produce compounds inducing positive skin reaction, and a mixture of *Cladosporium cladosploides* and *Fusarium nivale* produced ruminal atony in sheep. Later surveys[181,182] revealed that *F. nivale* produced a toxic butenolide with a less amount of T-2 toxin. The structure of butenolide was elucidated to be 4-acetoamide-4-hydroxybutenoic acid γ-lactone (Figure 1).

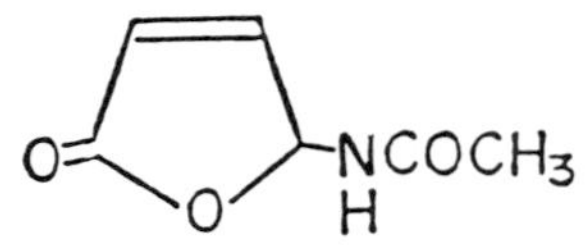

FIGURE 2. Structure of butenolide.

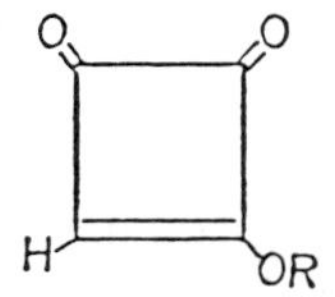

FIGURE 3. Structure of moniliformin (R = Na or K).

The dermal toxin, butenolide, was also reported to be produced by *F. equiseti,*[183] *F. graminearum,*[184] and *F. tricinctum.*[185] During the course of toxicological study on the metabolites of *F. sporotrichioides* NRRL 3510, one of the causative fungi of ATA, the authors isolated a large amount of butenolide along with T-2 toxin and related trichothecenes.[186]

Butenolide has a LD_{50} in mice of 43.6 mg/kg (i.p.) and 275 mg/kg (p.o.).[187] In the author's laboratory, in vivo and in vitro toxicities of butenolide were studied. The LD_{50} of the toxin in male mice was 91 mg/kg (i.p.); the lethal cases were observed mostly within 48 hr after administration. Pathologically, cellular degeneration and karyorrhexis were observed in the spleen, thymus, and lymph nodes. Since this toxin possesses *N*-acetylamino group in its molecule, the formation of methemoglobin in vivo was examined in the author's laboratory.[188] The toxin induced the increment of methemoglobin at concentrations of 1 to 3 mg/mℓ, when red blood cells were incubated at 37°C in vitro as *p*-aminophenol did. An interesting fact is that the toxin showed a latent period of 2 to 3 hr after the addition to blood cells for induction of methemoglobin, while *p*-aminophenol induced it immediately after the addition. This means that butenolide exhibits its activity after transformation to more reactive metabolite(s). These findings suggest that butenolide is one of the red blood toxins, and the impairment of oxygen transport in blood circulation is presumed to cause a decrease in tail temperature and gangrene in the intoxicated animals. Actually, in cattle injected with 3.8 mg/kg of the toxin (i.m.) for 90 days, dry gangrene at the end of the tail developed, and this lesion was one of the most characteristic signs observed in cattle grazing on tall fescue grass.[189]

In chickens which received T-2 toxin and butenolide together, mucosal necrosis in the mouth and crop, atrophy of the bursa fabricius and of the thymus, and an inflammation of the intesinal mucosa were evident.[190] However, these clinical changes were reported as a typical sign of T-2 toxin and related trichothecenes, and therefore it is difficult to evaluate these data in connection with the toxicity of butenolide.

Recently, Burmeister et al.[191] examined the long-term toxicity of butenolide in mice. The toxin was dissolved in water at concentrations of 0.1 and 0.5 mg/mℓ and was given to mice for 21 days. No significant effect was observed on body weight gain or survival, even though the daily amount consumed was near or exceeded the oral LD_{50}. The data indicated that the mouse can readily excrete or rapidly detoxify the toxin.

V. MONILIFORMIN

Moniliformin is a toxic metabolite of *F. moniliforme*. The southern corn leaf blight epidemic in the U.S. in 1970 seriously reduced corn yields in the Southeast and Midwest.

A toxigenic fungus identified as *F. moniliforme* was isolated and a water-soluble toxin, named moniliformin, was isolated as its metabolite.[192] X-ray crystallographic analysis revealed its structure as potassium salt of 1-hydroxycyclobut-1-ene-3,4-dione (Figure 2).[193] As producers of this toxin, *F. fusarioides, F. moniliforme* var. *subglutinans,* and *F. avenaceum* have been reported.[194,195]

The LD_{50} of moniliformin was 3.68 mg/kg in 7-day-old Peking duckling, and 41.6 and 50.0 mg/kg in female and male rats, respectively. A rapid progressive muscular weakness, respiratory stress, cyanosis, terminal coma, and death are the major symptoms of intoxication. Autopsy of the rats revealed acute congestive heart failure and focal myocardial degeneration and necrosis, swelling and scattered cell necrosis in the liver, kidney, pancrease, adrenal glands, gastric mucosal glands, and crypt of the small intestine.[195]

Burmeister et al.[192,196] observed the toxicity of moniliformin in chick embryos, chicks, and mice. The LD_{50} of the toxin in the embryo was 2.8 μg per egg, and those in chicks given by crop intubation and in female and male mice injected i.p. were 5.4, 20.9, and 29.1 mg/kg, respectively. Subchronic toxicity testing revealed that mice survived 20 days without apparent illness while consuming the oral LD_{50}.

Because of acute degenerative lesions in the heart and other organs, an involvement of supression of ATP-derived transmembrane transport mechanism is suspected. Thiel[197] investigated the mode of action of moniliformin. The results indicated that this toxin is a potent inhibitor of mitochondrial pyruvate and α-ketoglutarate oxydation, and the IC_{50} was estimated to be 5 μM. The enzymic systems responsible for the oxidative decarboxylation of pyruvate to acetyl-CoA and that of α-ketoglutarate to succinyl-CoA are similar. Both reactions consist of a multi-enzyme complex containing three different enzymes and five coenzymes, and they are sensitive to arsenate and arsenicals through the inactivation of lipoates. The transfer of acyl groups to coenzymes is suppressed by these inhibitors. Since these metabolic inhibitors and moniliformin are similar, it is presumed that moniliformin exerts its inhibitory effect on the oxidative reactions in a manner similar to arsenicals. This means that the primary effect of moniliformin is the depression of ATP formation in the tissues, and the rapid lethal action of the toxin resembles the action of other biochemical inhibitors such as fluoroacetate and cyanide.

Fusarium moniliforme Sheldon (*Gibberella fujikuroi* Sawada in perfect stage) occurs world-wide on a great variety of plant hosts and is one of the most prevalent fungi associated with corn kernels in most corn-producing areas of the world. According to Marasas et al.,[198] this fungus has recently been shown to be the cause of equine leukoencephalomalacia. This is a highly fatal disease affecting horses and other equine, and is characterized by liquifactive necrotic lesions in the subcortical white matter of the cerebral hemispheres. The toxigenic *F. moniliforme* produces toxic metabolites to experimental animals such as mice, rats, and others, but typical brain lesions of the disease have not been reproducible in any other animal species except the equine. At present, the agent responsible for equine leuko-encephalomalacia has not been clarified.

Another problem is the association of toxicants of *F. moniliforme* with the development of esophageal cancer. In a comparative study of the mycoflora of home-grown corn produced in Transkei, South Africa, a significantly higher incidence of *F. moniliforme* was found in corn produced in an area with a high human esophageal cancer rate than in corn from an area with a low human esophageal cancer rate.[199] In China, *F. moniliforme* is also one of the fungi most frequently detected from foodstuffs in areas with a high risk for esophageal cancer.[200]

Recently, Gelderblom et al.[20] surveyed the mutagenic metabolite *F. moniliforme* by employing the *Salmonella* mutagenicity assay. The main mutagenic compound purified from *F. moniliforme* seems to be identical with fusarin C.[202]

REFERENCES

1. **Godtfredsen, W. O., Grove, J. F., and Tamm, C.,** Zur Nomenklatur einer neuen Klasse von Sesquiterpen, *Helv. Chim. Acta,* 50, 1666, 1967.
2. **Ueno, Y., Sato, N., Ishii, K., Tsunoda, H., and Enomoto, M.,** Biological and chemical detection of trichothecene mycotoxins of *Fusarium* species, *Appl. Microbiol.,* 25, 699, 1973.
3. **Brian, P. W., Dawkins, A. W., Grove, J. F., Hemming, H. G., Lowe, D., and Norris, G. L. F.,** Phytotoxic compounds produced by *Fusarium equiseti, J. Exp. Bot.,* 12, 1, 1961.
4. **Saito, M. and Ohtsubo, K.,** Trichothecene toxins of *Fusarium* species, in *Mycotoxins,* Purchase, I. F. H., Ed., Elsevier, Amsterdam, 1974, 263.
5. **Ueno, Y., Ueno, I., Iitoi, Y., Tsunoda, H., Enomoto, M., and Ohtsubo, K.,** Toxicological approaches to the metabolites of *Fusaria.* III. Acute toxicity of fusarenon-X, *Jpn. J. Exp. Med.,* 41, 521, 1971.
6. **Matsuoka, Y., Kubota, K., and Ueno, Y.,** General pharmacological studies of fusarenon-X, a trichothecene mycotoxin from *Fusarium* spp., *Toxicol. Appl. Pharmacol.,* 50, 87, 1979.
7. **Matsuoka, Y. and Kubota, K.,** Studies on mechanisms of diarrhea induced by fusarenon-X, a trichothecene mycotoxin from *Fusarium* species, *Toxicol. Appl. Pharmacol.,* 57, 293, 1981.
8. **Sato, N., Ueno, Y., and Enomoto, M.,** Toxicological approaches to the metabolites of *Fusaria.* VIII. Acute and subacute toxicities of T-2 toxin in cats, *Jpn. J. Pharmacol.,* 25, 263, 1975.
9. **Yoshizawa, T. and Morooka, N.,** Trichothecenes from mold-infested cereals in Japan, in *Mycotoxins in Human and Animal Health,* Rodricks, J. V., Hesseltine, C. W., and Mehlman, A. M., Eds., Chem-Orbital Park Forest South, 1977, 309.
10. **Vesonder, R. F., Ciegler, A., and Jensen, A. H.,** Isolation of the emetic principle from *Fusarium*-infected corn, *Appl. Microbiol.,* 26, 1008, 1973.
11. **Ueno, Y., Ishii, K., Sato, N., and Ohtsubo, K.,** Toxicological approaches to the metabolites of *Fusaria.* VI. Vomiting factor from moldy corn infected with *Fusarium* spp., *Jpn. J. Exp. Med.,* 44, 123, 1974.
12. **Goodwin, W., Haas, C. D., Fabian, C., Heller-Bettinger, I., and Hoogstraten, B.,** Phase I evaluation of anguidine, *Cancer,* 42, 23, 1978.
13. **Chi, M. S., El-Halawant, M. E., Weibbel, P. E., and Mirocha, C. J.,** Effect of T-2 toxin on brain catecholamines and selected blood components in growing chickens, *Poultry Sci.,* 60, 137, 1981.
14. **Bamburg, J. R., Riggs, N. V., and Strong, F. M.,** The structures of toxins from two strains of *Fusarium tricinctum, Tetrahedron,* 24, 3329, 1968.
15. **Bamburg, J. R. and Strong, F. M.,** Mycotoxins of the trichothecene family produced by *Fusarium tricinctum* and *Trichoderma lignorum, Phytochemistry,* 8, 2405, 1969.
16. **Ueno, Y., Ishikawa, Y., Amakai, K., Nakajima, M., Saito, M., Enomoto, M., and Ohtsubo, K.,** Comparative study on skin-necrotizing effect of scirpene metabolites of *Fusaria, Jpn. J. Exp. Med.,* 40, 33, 1970.
17. **Hayes, M. A. and Schiefer, H. B.,** Quantitative and morphological aspects of cutaneous irritation by trichothecene mycotoxins, *Food Cosmet. Toxicol.,* 17, 611, 1979.
18. **Ueno, Y.,** Toxicological evaluation of trichothecene mycotoxins, in *Natural Toxins,* Eaker, D. and Wedstrom, T., Eds., Pergamon, New York, 1980, 663.
19. **Smalley, E. B., Marasas, W. F. O., Strong, F. M., Bamburg, J. R., Nichols, R. E., and Kosuri, N. R.,** Mycotoxicoses associated with moldy corn, in *Toxic Microorganisms,* Herzberg, M., Ed., UJNR Joint Panel on Toxic Microorganisms and the U.S. Department of the Interior, Washington, D.C., 1970, 163.
20. **Sato, N. and Ueno, Y.,** Comparative toxicity of trichothecenes, in *Mycotoxins in Human and Animal Health,* Rodricks, J. V., Hesseltine, C. W., and Mehlman, M. A., Eds., Chem-Orbital, Park Forest South, 1977, 295.
21. **Sato, N., Ito, T., Kumada, H., Ueno, Y., Asano, K., Saito, M., Ohtubo, K., Ueno, I., and Hatanaka, Y.,** Toxicological approaches to the metabolites of *Fusaria.* XIII. Hematological changes in mice by a single and repeated administrations of trichothecenes, *J. Toxicol. Sci.,* 3, 335, 1978.
22. **Wyatt, R. D., Doerr, J. A., Hamilton, P. B., and Burmeister, H. R.,** Egg production shell thickness and other physiological parameters of laying hens affected by T-2 toxin, *Appl. Microbiol.,* 29, 641, 1975.
23. **Chi, M. S., Mirocha, C. J., Kurz, H. J., Weaver, G., Bates, F., and Shimoda, W.,** Subacute toxicity of T-2 toxin in broiler chicks, *Poultry Sci.,* 56, 306, 1977.
24. **Richard, J. L., Cysewski, S. J., Pier, A. C., and Booth, G. D.,** Comparison of effects of dietary T-2 toxin on growth, immunogenic organs, antibody formation, and pathologic changes in turkeys and chickens, *Am. J. Vet. Res.,* 39, 1647, 1978.
25. **DeNicola, D. B., Reber, A. H., Carlton, W. W., and Yagen, B.,** T-2 toxin mycotoxicosis in guinea pig, *Food Cosmet. Toxicol.,* 63, 2015, 1979.
26. **Rukmini, C., Prasad, J. S., and Rao, K.,** Effect of feeding of T-2 toxin to rats and monkeys, *Food Cosmet. Toxicol.,* 18, 267, 1980.
27. **Stahelin, J. P., Kalbered-Rusch, M. E., Singer, E., and Alizary, S.,** Some biological effects of the cytostic diacetoxyscirpenol, *Arzneim. Forsch.,* 18, 989, 1968.

28. **Ueno, Y.,** Trichothecene mycotoxins: mycology, chemistry, and toxicology, in *Advance in Nutritional Research,* Vol. 3, Draper, H. H., Ed., Plenum Press, New York, 1980, 301.
29. **Segal, R., Milo-Goldzweig, L., Joffe, A. Z., and Yagen, B.,** Trichothecene-induced hemolysis. I. The hemolytic activity of T-2 toxin, *Toxicol. Appl. Pharmacol.,* 70, 343, 1983.
30. **Ueno, Y. and Kubota, K.,** DNA-attacking ability of carcinogenic mycotoxins in recombination deficient mutant cells of *Bacillus subtilis, Cancer Res.,* 36, 445, 1976.
31. **Ueno, Y., Kubota, K., Ito, T., and Nakamura, Y.,** Mutagenicity of carcinogenic mycotoxins in *Salmonella typhimurium, Cancer Res.,* 38, 536, 1978.
32. **Kuczuk, M. H., Benson, P. M., Heath, H., and Hayes, A. W.,** Evaluation of the mutagenic potential of mycotoxins using *Salmonella typhimurium* and *Saccharomyces cerevisiae, Mutat. Res.,* 53, 11, 1978.
33. **Wehner, F. C., Marasas, W. F. O., and Thiel, P. G.,** Lack of mutagenicity to *Salmonella typhimurium* of some *Fusarium* myocotoxins, *Appl. Environ. Microbiol.,* 35, 659, 1978.
34. **Umeda, M., Tsutsui, T., and Saito, M.,** Mutagenicity and inducibility of DNA single-strand breaks and chromosome aberrations by various mycotoxins, *Gann,* 68, 619, 1977.
35. **Norppa, H., Penttila, M., Sorsa, M., Hintika, E. L., and Ilus, T.,** Mycotoxin T-2 of *Fusarium tricinctum* and chromosome changes in Chinese hamster bone marrow, *Hereditas,* 93, 329, 1980.
36. **Lafarge-Frayssinet, C., Decloitre, F., Mousset, S., Martin, M., and Frayssinet, C.,** Induction of DNA single strand breaks by T-2 toxin, a trichothecene metabolite of *Fusarium,* effect on lymphoid organs and liver, *Mutat. Res.,* 88, 115, 1981.
37. **Reiss, J.,** Mycotoxin poisoning of *Allium cepa* root tips. II. Reduction of mitotic index and formation of chromosomal aberrations and cytological abnormalities by paturin, rubratoxin B and diacetoxyscirpenol, *Cytologia,* 39, 703, 1975.
38. **Stanford, G. K., Hood, R. D., and Hayes, A. W.,** Effect of prenatal administration of T-2 toxin to mice, *Res. Commun. Chem. Pathol. Pharmacol.,* 10, 743, 1975.
39. **Ito, Y., Ohtsubo, K., and Saito, M.,** Effect of fusarenon-X a trichothecene produced by *Fusarium nivale* on prenant mice and their fetus, *Jpn. J. Exp. Med.,* 50, 167, 1980.
40. **Khera, K. S., Whalen, C., Angers, G., Vesonder, R. F., and Kuiper-Goodman, T.,** Embryotoxicity of 4-deoxynivalenol (vomitoxin) in mice, *Bull. Environ. Contam. Toxicol.,* 29, 487, 1982.
41. **Marasas, W. F. O., Bamburg, J. R., Smalley, E. B., Strong, F. M., Regland, W. L., and Degurse, P. E.,** Toxic effect on trout, rat and mice of T-2 toxin produced by the fungus *Fusarium tricinctum, Toxicol. Appl. Pharmacol.,* 15, 471, 1969.
42. **Ohtsubo, K. and Saito, M.,** Chronic effect of trichothecene toxins, in *Mycotoxin in Human and Animal Health,* Rodricks, J. V., Hesseltine, C. W., and Mehlman, A. M., Eds., Chem-Orbital, Park Forest South, 1977, 255.
43. **Schoental, R., Joffe, A. Z., and Yagan, B.,** Cardiovascular lesions and various tumors found in rats given T-2 toxin, a trichothecene metabolite of *Fusarium, Cancer Res.,* 39, 2179, 1979.
44. **Saito, M., Horiuchi, T., Ohtsubo, K., Hatanaka, Y., and Ueno, Y.,** Low tumor incidence in rats with long-term feeding of fusarenon-X, a cytotoxic trichothecene produced by *Fusarium nivale, Jpn. J. Exp. Med.,* 50, 293, 1980.
45. **Ueno, Y. and Shimada, Y.,** Unpublished data, 1979.
46. **Lindenfelser, L. A., Lellehoj, E. B., and Burmeister, H. R.,** Aflatoxin and trichothecene toxins: skin tumor induction and synergistic acute toxicity in white mice, *J. Natl. Cancer Inst.,* 52, 113, 1974.
47. **Reiss, J.,** Influence of the mycotoxin paturin and diacetoxyscirpenol on fungi, *J. Gen. Appl. Microbiol.,* 19, 415, 1973.
48. **Bamburg, J. R., Marasas, W. F. O., Smalley, E. B., and Strong, F. M.,** Toxic compounds from *Fusaria* and other *Hyphomycetes, Biotech. Bioeng.,* 10, 445, 1968.
49. **Bamburg, J. R. and Strong, F. M.,** 12,13-Epoxytrichothecenes, in *Microbial Toxins,* Vol. 7, Kadis, S., Ciegler, A., and Ajl, S. J., Eds., Academic Press, New York, 1971, 207.
50. **Burmeister, H. R. and Hesseltine, C. W.,** Biological assay for two mycotoxins produced by *Fusarium tricinctum, Appl. Microbiol.,* 20, 437, 1970.
51. **Marasas, W. F. O., Smalley, E. B., Bamburg, J. R., and Strong, F. M.,** Phytotoxicity of T-2 toxin produced by *Fusarium tricinctum, Phytopathology,* 61, 1488, 1971.
52. **Helgeson, J. P., Haberach, G. T., and Vanderhoef, L. N.,** T-2 toxin decreases logarithmic growth rates of tobacco callus tissues, *Plant Physiol.,* 52, 660, 1973.
53. **Siriwardamp, T. M. G. and Lafont, P.,** New sensitive biological assay for 12,13-epoxytrichothecenes, *Appl. Environ. Microbiol.,* 35, 206, 1978.
54. **Stahl, C., Vanderhoef, L. N., Siegel, N., and Helgeson, J. P.,** Fusarium tricinctum T-2 toxin inhibits auxin promoted elongation in soybean hypocotyl, *Plant Physiol.,* 52, 663, 1973.
55. **Ueno, Y. and Yamakawa, H.,** Antiprotozoal activity of scirpen mycotoxins of *Fusarium nivale* Fn 2B, *Jpn. J. Exp. Med.,* 40, 385, 1970.
56. **Dive, D., Moreau, S., and Cacan, M.,** Use of a cilliate protozoan for fungal toxin studies, *Bull. Environ. Contam. Toxicol.,* 19, 489, 1978.

57. **Brown, R. F., Wildan, J. P., and Eppley, R. M.,** Temperature-dose relationships with aflatoxin on the brine shrimp, *Artemia salina, J. Assoc. Off. Anal. Chem.,* 51, 905, 1968.
58. **Harwig, J. and Scott, P. M.,** Brine shrimp *(Artemia salina)* larvae as a screening system for fungal toxins, *Appl. Microbiol.,* 21, 1011, 1971.
59. **Siegfried, R. and Frank, H. K.,** A contribution to the analysis of *Fusarium* toxins (trichothecene toxins), *Z. Lebensm. Unters. Forsch.,* 166, 363, 1978.
60. **Reiss, J.,** Comparative investigations on the toxicity of some mycotoxins to the larvae of the brine shrimp *(Artemia salina), Zentralbl. Bakteriol. Parasitenk. Infektionskr. Hyg.,* 2, 155, 1972.
61. **Eppley, R. M.,** Sensitivity of brine shrimp *(Artemia salina)* to trichothecenes, *J. Assoc. Off. Anal. Chem.,* 57, 618, 1974.
62. **Osame, J., Yoshizawa, T., Kurokawa, Y., and Morooka, N.,** Toxicity of mycotoxins as assayed by the early developmental sea urchin embryo, *Proc. Jpn. Assoc. Mycotoxicol.,* 8, 22, 1978.
63. **Grove, J. F. and Hosken, M.,** Larvicidal activity of some 12,13-epoxytrichothec-9-enes, *Biochem. Pharmacol.,* 24, 959, 1975.
64. **Wright, V. F., Casas, E. D. L., and Harein, P. K.,** The response of *Tribolium confusum* to the mycotoxins zearalenone F-2 and T-2 toxin, *Environ. Entomol.,* 5, 321, 1976.
65. **Reiss, J.,** Toxic effect of the mycotoxin, aflatoxin B_1, rubratoxin B, paturin, and diacetoxyscirpenol on the crustacean *Cyclops fuscus, J. Assoc. Off. Anal. Chem.,* 55, 895, 1972.
66. **Abedi, Z. H. and Scott, P. M.,** Detection of toxicity of aflatoxins, sterigmatocystin, and other fungal toxins by lethal action of zebra fish larvae, *J. Assoc. Off. Anal. Chem.,* 52, 963, 1969.
67. **Ichinoe, M., Kurata, H., and Ueno, Y.,** Researches on toxic fungi in foodstuff. XVI. Chick embryo method for screening of toxic metabolites of *Fusarium* spp, *Jpn. J. Food Hyg. Sci.,* 31, 481, 1972.
68. **Reiss, J.,** Toxic effects of mycotoxins on amphibian larvae, *Toxicology,* 8, 121, 1977.
69. **Grove, J. F. and Mortimer, P. H.,** The cytotoxicity of some transformation products of diacetoxyscirpenol, *Biochem. Pharmacol.,* 18, 1473, 1969.
70. **Ohtsubo, K., Yamada, M., and Saito, M.,** Inhibitory effect of nivalenol, a toxic metabolite of *Fusarium nivale,* on the growth cycle and biopolymer synthesis of Hela cells, *Jpn. J. Med. Sci. Biol.,* 21, 185, 1968.
71. **Ohtsubo, K. and Saito, M.,** Cytotoxic effects of scirpen compounds fusarenon-X produced by *Fusarium nivale,* dihidronivalenol and dihydrofusarenon-X on Hela cells, *Jpn. J. Med. Sci. Biol.,* 23, 217, 1970.
72. **Boden, L. and Zoldag, L.,** Cytotoxicity studies on T-2 fusariotoxin, *Acta Vet. Acad. Sci. Hung.,* 24, 451, 1974.
73. **Tanaka, T., Matsuda, Y., Toyazaki, N., Ogawa, K., Matsuki, Y., and Ueno, Y.,** Screening of trichothecene-producing *Fusarium* species from river sediments by mammalian cell culture techniques, *Proc. Jpn. Assoc. Mycotoxicol.,* 5/6, 50, 1977.
74. **Mirocha, C. J.,** Trichothecene toxins produced by *Fusarium,* in *Proc. Conference on Mycotoxins in Animal Feeds and Grains to Animal Health,* 1979, 288.
75. **Rosenstein, Y., Lafarge-Frayssinet, C., Lespinate, G., Loisillier, F., Lafont, P., and Frayssinet, C.,** Immunosupressive activity of *Fusarium* toxins effects on antibody synthesis and skin grafts of crude extracts T-2 toxin and diacetoxyscirpenol, *Immunology,* 36, 111, 1979.
76. **Masuko, H., Ueno, Y., Otokawa, M., and Yaginuma, K.,** The enhancing effect of T-2 toxin on delayed hypersensitivity in mice, *Jpn. J. Med. Sci. Biol.,* 30, 159, 1977.
77. **Otokawa, M., Shibata, Y., and Egashira, Y.,** The Inhibitory effect of T-2 toxin on tolerance induction of delayed type hyper-sensitivity in mice, *Jpn. J. Med. Sci. Biol.,* 32, 37, 1979.
78. **Masuda, E., Takemoto, T., Tatsuno, T., and Obara, T.,** Immunosupressive effect of a trichothecene mycotoxin, Fusarenon-X in mice, *Immunology,* 45, 743, 1982.
79. **Sato, S., Ohya, T., Homma, S., and Tatsuno, T.,** Effect of fusarenon-X on serological responses of chicks inoculated with newcastle disease vaccines, *Proc. Jpn. Assoc. Mycotoxicol.,* 13, 43, 1981.
80. **Lafarge-Frayssinet, C., Lespinants, G., Lafornt, P., Loisillier, F., Mousset, S., Rosenstein, Y., and Frayssinet, C.,** Immunosupressive effects of *Fusarium* extracts and trichothecene on blastogenic response of murine splenic and thymic cells to mitogen, *Proc. Soc. Exp. Biol. Med.,* 160, 302, 1979.
81. **Boonchuvit, B., Hamilton, P. B., and Burmeister, H. R.,** Interaction of T-2 toxin with *Salmonella* infections of chickens, *Poultry Sci.,* 54, 1693, 1975.
82. **Kanai, K. and Kondou, E.,** Effect of T-2 toxin on the resistance to infection, in *Abstr. 19th Meet. Jpn. Assoc. Mycotoxicol.,* Abstr. 4, 1984.
83. **Ueno, Y., Hosoya, M., Morita, Y., Ueno, I., and Tatsuno, T.,** Inhibition of the protein synthesis in rabbit reticulocytes by nivalenol, a toxic principle isolated from *Fusarium nivale* growing on rice, *J. Biochem. (Tokyo),* 64, 479, 1968.
84. **Ueno, Y. and Fukushima, M.,** Inhibition of protein synthesis and DNA synthesis in Ehrlich ascites tumor by nivalenol, a toxic principle of *Fusarium nivale* growing rice, *Experientia,* 24, 1032, 1968.
85. **Ueno, Y.,** Inhibition of protein synthesis in animal cells by nivalenol and related metabolites: toxic principle of rice infested with *Fusarium nivale,* in *Toxic Micro-Organisms,* Herzberg, M., Ed., UJNR Joint Panel on Toxic Microorganisms and the U.S. Department of the Interior, Washington, D.C., 1970, 76.

86. **Liao, L. L., Grollman, A. P., and Horwits, S. B.,** Mechanism of action of the 12,13-epoxytrichothecene anguidine: an inhibitor of protein synthesis, *Biochim. Biophys. Acta,* 454, 273, 1976.
87. **Ueno, Y., Nakajima, M., Sakai, K., Ishii, K., Sato, N., and Shimada, N.,** Comparative toxicology of trichothecene mycotoxins: inhibition of protein synthesis in animal cells, *J. Biochem. (Tokyo),* 74, 285, 1973.
88. **Gupta, R. S. and Siminovitch, L.,** Genetic and biochemical characterization of mutants of CHO cells resistant to the protein synthesis inhibitor trichodermin, *Somatic Cell Genet.,* 4, 355, 1978.
89. **Carasco, L., Barbacid, M., and Vazquez, D.,** The trichodermin group of antibiotics inhibitors of peptide bond formation by eukaryotic ribosomes, *Biochim. Biophys. Acta,* 312, 368, 1973.
90. **Stafford, M. E. and McLaughlin, C. S.,** Trichodermin, a possible inhibitor of the termination process of protein synthesis, *J. Cell Physiol.,* 82, 121, 1973.
91. **Cundliffe, E., Cannon, M., and Davis, J.,** Mechanism of inhibition of eukaryotic protein synthesis by trichothecene fungal toxins, *Proc. Natl. Acad. Sci. U.S.A.,* 71, 30, 1974.
92. **McLaughlin, C. S., Vaughamm, M. H., Cambell, I. M., Wei, C. M., Stafford, M. E., and Hansen, B. S.,** Inhibition of protein synthesis by trichothecenes, in *Mycotoxin in Human and Animal Health,* Rodricks, J. V., Hesseltine, C. W., and Mehlman, M. A., Eds., Chem-Orbital, Park Forest South, 1977, 263.
93. **Ueno, Y., Hosoya, M., and Ishikawa, Y.,** Inhibitory effect of mycotoxins on protein synthesis in rabbit reticulocytes, *J. Biochem. (Tokyo),* 66, 419, 1969.
94. **Ueno, Y. and Shimada, N.,** Reconfirmation of the specific nature of reticulocytes bioassay system to trichothecene mycotoxins of *Fusarium* spp., *Chem. Pharm. Bull.,* 22, 2744, 1974.
95. **Wei, C. M. and McLaughlin, C. S.,** Structure function relationship in the 12,13-epoxytrichothecenes novel inhibitors of protein synthesis, *Biochem. Biophys. Res. Commun.,* 57, 838, 1974.
96. **Carter, C. J., Cannon, M., and Smith, K. E.,** Inhibition of protein synthesis in reticulocyte lysates by trichodermin, *Biochem. J.,* 154, 171, 1976.
97. **Mizuno, S.,** Mechanism of inhibition of protein synthesis initiation by diacetoxyscirpenol and fusarenon-X in the reticulocyte lysate system, *Biochim. Biophys. Acta,* 383, 207, 1975.
98. **Hartman, G. R., Richter, H., Weiner, E. M., and Zimmermann, W.,** On the mechanism of action of the cytostatic drug anguidine and of the immunosupressive agent ovalcin, two sesquiterpenes from fungi, *Planta Med.,* 34, 231, 1978.
99. **Ohtsubo, K., Kaden, P., and Mittermeyer, C.,** Polyribosomal breakdown in mouse fibroblast L-cells by fusarenon-X, a toxic principle isolated from *Fusarium nivale, Biochim. Biophys. Acta,* 287, 520, 1972.
100. **Schindler, D.,** Two classes of inhibitors of peptidyl transferase activity in eukaryotes, *Nature (London),* 249, 38, 1974.
101. **Tate, W. P. and Caskey, C. T.,** Peptidyl transferase inhibition by trichodermin, *J. Biol. Chem.,* 248, 7970, 1973.
102. **Smith, K. E., Cannon, M., and Cundliffe, E.,** Inhibition at the initiation level of eukaryotic protein synthesis by T-2 toxin, *FEBS Lett.,* 50, 8, 1975.
103. **Wei, C. M., Campbell, I. M., McLaughlin, C. S., and Maurice, H.,** Binding of trichodermin to mammalian ribosomes and its inhibition by other 12,13-epoxytrichothecenes, *Mol. Cell. Biochem.,* 3, 215, 1974.
104. **Cannon, M., Smith, K. E., and Carter, C. J.,** Prevention by ribosome bound nascent polyphenylalanine chains of the functional interaction of T-2 toxin with its receptor site, *Biochem. J.,* 156, 289, 1976.
105. **Jimenez, A. and Vasquez, D.,** Quantitative binding of antibiotics to ribosomes from a yeast mutant altered on the peptidyl transferase center, *Eur. J. Biochem.,* 54, 483, 1975.
106. **Schindler, D., Grant, P., and Davies, J.,** Trichodermin resistant mutation affecting eukaryotic ribosomes, *Nature (London),* 248, 535, 1974.
107. **Jimenez, A., Sanchez, L., and Vazquez, D.,** Simultaneous ribosomal resistance to trichodermin and anisomycin in *Saccharomyces cerevisiae* mutants, *Biochim. Biophys. Acta,* 383, 427, 1975.
108. **Grant, P. G., Schindler, D., and Davies, J. E.,** Mapping of trichodermin resistance in *Saccharomyces cerevisiae:* a genetic locus for a component of the 60S ribosomal subunit, *Genetics,* 83, 667, 1976.
109. **Carter, C. J., Cannon, M., and Jimenet, A.,** A trichodermin resistant mutant of *Saccharomyces cerevisiae* with an abnormal distribution of native ribosomal subunits, *Eur. J. Biochem.,* 107, 173, 1980.
110. **Fried, H. M. and Waner, J. R.,** Cloning of yeast gene for trichodermin resistance and ribosomal protein L3, *Proc. Natl. Acad. Sci. U.S.A.,* 78, 238, 1981.
111. **Hobden, A. N. and Cundliffe, E.,** Ribosomal resistance to the 12,13-epoxytrichothecene antibiotics in the producing organism *Myrothecium verrucaria, Biochem. J.,* 190, 765, 1980.
112. **Sun, S. M. M., Buchbinder, B. U., and Hall, T. C.,** Cell-free synthesis of the major storage protein of the bean *Phaseolus vulgaris, Plant Physiol.,* 56, 780, 1975.
113. **Tashiro, F., Hirai, K., and Ueno, Y.,** Inhibitory effects of carcinogenic mycotoxins on deoxyribonucleic acid-dependent ribonucleic acid polymerase and polyribonuclease H, *Appl. Environ. Microbiol.,* 38, 191, 1979.

114. **Munsch, N. and Mueller, W. E. G.,** Effect of T-2 toxin on DNA polymerases and terminal deoxynucleotidyl transferase of Molt-4 and Nu-8 cell lines, *Immunopharmacology,* 2, 313, 1980.
115. **Agrelo, C. E. and Schoental, R.,** Synthesis of DNA in human fibroblasts treated with T-2 toxin and HT-2 toxin (the trichothecene metabolites of *Fusarium* species) and the effects of hydroxyurea, *Toxicol. Lett.,* 5, 155, 1980.
116. **Iwahashi, T., Tashiro, F., and Ueno, Y.,** Mechanism of cytotoxic effect of T-2 toxin on a protozoan, *Tetrahymena pyriformis* GL, *Proc. Jpn. Assoc. Mycotoxicol.,* 15, 31, 1982.
117. **Chiba, J., Nakano, N., Morooka, N., Nakazawa, S., and Watanabe, Y.,** Inhibitory effect of fusarenon-X, a sesquiterpene mycotoxin on lipid synthesis and phosphate uptake in *Tetrahymena pyriformis, Jpn. J. Med. Sci. Biol.,* 25, 291, 1972.
118. **Ueno, Y., Matsumoto, H., Ishii, K., and Kukita, K.,** Inhibitory effects of mycotoxins on Na^+ dependent transport of glycine in rabbit reticulocytes, *Biochem. Pharmacol.,* 25, 2091, 1978.
119. **Ueno, Y. and Matsumoto, H.,** Inactivation of some thiol enzyme by trichothecene mycotoxins from *Fusarium* spp., *Chem. Pharm. Bull.,* 23, 2439, 1975.
120. **Reiss, J.,** Inhibition of urease by the mycotoxin paturin, *Naturwissenschaften,* 64, 97, 1977.
121. **Farb, R. M., Mego, J. L., and Hayes, A. W.,** Effect of mycotoxin on uptake and degradation of iodine-125 albumin in mouse liver and kidney lysosomes, *J. Toxicol. Environ. Health,* 1, 985, 1976.
122. **Foster, P. M. D., Slater, T. F., and Patterson, D. S. P.,** A possible enzymic assay for trichothecene mycotoxins in animal feedstuffs, *Biochem. Soc. Trans.,* 3, 875, 1975.
123. **Coffin, J. L. and Comb, G. F., Jr.,** Impaired vitamin E status of chicks fed T-2 toxin, *Poultry Sci.,* 60, 385, 1981.
124. **Shimizu, T., Nakano, N., Matsui, T., and Aibara, K.,** Hypoglycemia in mice administered fusarenon-X, *Jpn. J. Med. Sci. Biol.,* 32, 189, 1979.
125. **McNutt, S. H., Ourwin, P., and Murray, C.,** Vulvovaginitis in swine, *J. Am. Vet. Med. Assoc.,* 73, 484, 1928.
126. **McErlean, B. A.,** Vulvovaginitis of swine, *Vet. Record,* 64, 539, 1952.
127. **Dorfman, R. I. and Dorfman, A. S.,** *Proc. Animal Care Panel (New York),* 5, 16, 1955.
128. **Stob, M., Baldwin, R. S., Tuite, J., Andrews, F. N., and Gillette, K. G.,** Isolation of an anabolic uterotropic compound from corn infected with *Gibberella zeae, Nature (London),* 196, 1318, 1962.
129. **Christensen, C. M., Nelson, G. H., and Mirocha, C. J.,** Effect on the white rat uterus of a toxic substance isolated from *Fusarium, Appl. Microbiol.,* 13, 653, 1965.
130. **Urry, W. H., Wehrmister, H. L., Hodge, E. B., and Hidey, P. H.,** The structure of zearalenone, *Tetrahedron Lett.,* 27, 3109, 1966.
131. **Kuo, C. H., Taub, D., Hoffsommer, R. D., Wendler, N. L., Urry, W. H., and Mullenbach, G.,** Resolution of (±)-zearalenon. Determination of the absolute configuration of the natural enantiomorph, *Chem. Commun.,* 15, 761, 1967.
132. **Taub, D., Biorotra, N. N., Hoffsomer, R. D., Kuo, C. H., Slates, H. L., Weber, S., and Wendler, N. L.,** Total synthesis of the macrolide, zearalenone, *Tetrahedron,* 24, 2443, 1968.
133. **Giotra, N. N. and Wendler, N. L.,** A new total synthesis of the macrolide zearalenone, *Chem. Ind. (London),* 35, 1493, 1967.
134. **Giotra, N. N. and Wendler, N. L.,** Knoevenagel condensation in the homophtalic acid series. A synthesis of zearalenone, *J. Org. Chem.,* 34, 3192, 1969.
135. **Vlattas, I., Harrison, I. T., Tokes, L., Fried, J. H., and Cross, A. D.,** Synthesis of (±)-zearalenone, *J. Org. Chem.,* 33, 4176, 1968.
136. **Hurd, R. N. and Shah, D. H.,** Total synthesis of macrolide, (R,S)-zearalenone, *J. Med. Chem.,* 16, 543, 1973.
137. **Masamune, S., Kamata, S., and Schilling, W.,** Synthesis of macrolide antibiotics. III. Direct ester and lactone synthesis from S-*tert*-butyl thioate (thiol ester), *J. Am. Chem. Soc.,* 97, 3515, 1975.
138. **Mirocha, C. J., Christensen, C. M., and Nelson, G. H.,** Estrogenic metabolites produced by *Fusarium graminearum* in stored corn, *Appl. Microbiol.,* 15, 497, 1976.
139. **Jackson, R. A., Fenton, S. W., Mirocha, C. J., and Davis, G.,** Characterization of two isomers of 8′-hydroxyzearalenone and other derivatives of zearalenone, *J. Agric. Food Chem.,* 22, 1015, 1974.
140. **Steele, J. A., Mirocha, C. J., Pathre, S. V., and Behrens, J. C.,** Metabolism of zearalenone by *Fusarium roseum* Graminearum, *J. Agric. Food Chem.,* 24, 89, 1976.
141. **Hagler, W. M., Mirocha, C. J., Pathre, S. V., and Behrens, J. C.,** Identification of the naturally occurring isomer of zearalenol, *Appl. Microbiol.,* 37, 849, 1979.
142. **Bolliger, G. and Tamm, C.,** Vier neue Metabolite from *Gibberella zeae;* 5-Formylzearalenone, 7′-Dehydrozearalenone, 8′-Hydoxyzearalenone und 8′-Epi-Hydroxyzearalenone, *Helv. Chim. Acta,* 55, 3030, 1972.
143. **Ellastad, G. A., Lovelli, F. M., Perkinson, N. A., Hargreaves, R. T., and McGahren, W. J.,** New zearalenone related macrolides and isocoumarins from an unidentified fungus, *J. Org. Chem.,* 43, 2339, 1978.

144. **Hidy, P. H., Baldwin, R. S., Greasham, R. L., Keith, C. L., and McMullen, J. R.,** Zearalenone and some derivatives: production and biological activities, *Adv. Appl. Microbiol.*, 22, 59, 1977.
145. **Bailey, D. E., Cox, G. E., Morgareidge, K., and Taylor, J.,** Acute and subacute toxicity of zearalenone in the rat, *Toxicol. Appl. Pharmacol.*, 37, 144, 1976.
146. **Ruddick, J. A., Scott, P. M., and Harwig, J.,** Teratological evaluation of zearalenone administered orally to the rat, *Bull Environ. Contam.*, 15, 678, 1976.
147. **Ruzas, C., Mess, B., Biro-Gosztomyi, M., and Woller, L.,** in *Hormones and Brain Development*, Dorner, G. and Kawasaki, M., Eds., Elsevier, Amsterdam, 1978, 57.
148. **Vanyi, A. and Szailer, E.,** Investigation of the cytotoxic effect of F-2 toxin zearalenone in various monolayer cell cultures, *Acta Vet. Acad. Sci. Hung.*, 24, 407, 1974.
149. **Horakova, K. and Betina, V.,** Cytotoxic activity of macrocyclic metabolites from fungi, *Neoplasma*, 24, 21, 1977.
150. **Vianello, A. and Macri, F.,** Inhibition of plant cell membrane transport phenomena induced by zearalenone (F-2), *Planta*, 143, 51, 1978.
151. **Boutibonnes, P.,** Effects of zearalenone on the bacterial cell, *Mycopathology*, 69, 117, 1979.
152. National Cancer Institute, National Toxicology Program Technical Report on the Carcinogenesis Bioassay of Zearalenone, DHHS Publication No. (NHI) 81-1791, Bethesda, Md., 1981.
153. **Gibson, J. P., Newberne, J. W., Kuhn, W. L., and Elsea, J. R.,** Comparative chronic toxicity of three oral estrogens in rats, *Toxicol. Appl. Pharmacol.*, 11, 489, 1967.
154. WHO, IARC Monograph on the Evaluation of the Carcinogenic Risk of Chemicals to Human: Sex hormone (II), Vol. 21, World Health Organization, Lyon, France, 1979, 173.
155. **Ingerowski, G. H., Scheutwinkel-Reich, M., and Stan, H. J.,** Mutagenicity studies on veterinary anabolic drugs with the *Salmonella*/microsome test, *Mutat. Res.*, 91, 93, 1981.
156. **Tomatis, L., Agthe, C., Bartsch, H., Huff, J., Montesano, R., Sarcci, R., Walker, E., and Wilbourn, J.,** Evaluation of carcinogenicity of chemicals: a review of the monograph program of the international agency for research on cancer (1971 to 1977), *Cancer Res.*, 38, 877, 1978.
157. **Maugh, T. H.,** Chemical carcinogens: the scientific basis for regulation, *Science*, 201, 1200, 1978.
158. **Ueno, Y., Shimada, N., Yagasaki, S., and Enomoto, M.,** Toxicological approaches to the metabolites of *Fusaria*. VII. Effect of zearalenone on the uteri of mice and rats, *Chem. Pharm. Bull.*, 22, 2830, 1974.
159. **Mirocha, C. J., Christensen, C. M., and Nelson, G. H.,** Toxic metabolites produced by fungi implicated in mycotoxicoses, *Biotech. Bioeng.*, 10, 469, 1968.
160. **Hobson, W., Bailey, J., and Fuller, G. B.,** Hormone effects of zearalenone in nonhuman primates, *J. Toxicol. Environ. Health*, 3, 43, 1977.
161. **Ingerowski, G. H. and Stan, H. J.,** In vitro metabolism of the anabolic drug zeranol, *J. Environ. Pathol. Toxicol.*, 2, 1173, 1979.
162. **Greenman, D. L., Wittliff, J. L., Boyd, P. A., and Mheta, R. G.,** Interaction of *Fusarium* mycotoxins with estradiol binding sites in target tissues, *J. Toxicol. Environ. Health*, 3, 348, 1977.
163. **Greenman, D. L., Mheta, R. G., and Wittliff, J. L.,** Nuclear interaction of *Fusarium* mycotoxins with estradiol binding sites in the mouse uterus, *J. Toxicol. Environ. Health*, 5, 593, 1979.
164. **Boyd, P. A. and Wittliff, J. L.,** Mechanism of *Fusarium* mycotoxin action in mammary gland, *J. Toxicol. Environ. Health*, 4, 1, 1978.
165. **Kiang, D. T., Kennedy, B. J., Pathre, S. V., and Mirocha, C. J.,** Binding characteristics of zearalenone analogs to estrogen receptors, *Cancer Res.*, 38, 3611, 1978.
166. **Martin, P. M., Horwitz, K. B., Ryan, D. S., and McGuire, W. L.,** Phytoestrogen interaction with estrogen receptor in human breast cancer cells, *Endocrinology*, 103, 1860, 1978.
167. **Katzenellenbogen, B. S., Katzenellenbogen, J. A., and Mordecai, D.,** Zearalenone: characterization of the estrogenic potençies and receptor interactions of a series of fungal β-resorcilic acid lactones, *Endocrinology*, 105, 33, 1979.
168. **Verdeal, K., Brown, R. R., Richardson, T., and Ryan, D. S.,** Affinity of phytoestrogens for estradiol-binding proteins and effect of coumestrol on growth of 7,12-dimethylbenz(*a*)anthracene-induced rat mammary tumors, *J. Natl. Can. Res. Inst.*, 64, 285, 1980.
169. **Tashiro, F., Kawabata, Y., Naoi, M., and Ueno, Y.,** Zearalenone-estrogen receptor interaction and RNA synthesis in rat uterus, in *Medical Mycology*, Preusser, H. L., Ed., Gustav Fisher Verlag, Stuttgart, 1980, 311.
170. **Kitagawa, M., Tashiro, F., and Ueno, Y.,** Interaction between zearalenone, an estrogenic mycotoxin, and the estrogenreceptor of the rat brain, *Proc. Jpn. Assoc. Mycotoxicol.*, 15, 28, 1982.
171. **Ueno, Y. and Yagasaki, S.,** Toxicoligical approaches to the metabolites of *Fusaria*. X. Accelerating effect of zearalenone on RNA and protein synthesis in the uterus of ovariectomized mice, *Jpn. J. Exp. Med.*, 45, 199, 1975.
172. **Kawabata, Y., Tashiro, F., and Ueno, Y.,** Synthesis of a specific protein induced by zearalenone and its derivatives in rat uterus, *J. Biochem. (Tokyo)*, 91, 801, 1982.

173. **Ueno, Y., Ayaki, S., Sato, N., and Ito, T.,** Fate and mode of action of zearalenone, *Ann. Nutr. Aliment.*, 31, 935, 1977.
174. **Ueno, Y. and Tashiro, F.,** Alpha-zearalenol, a major hepatic metabolite in rats of zearalenone, an estrogenic mycotoxin of *Fusarium* species, *J. Biochem. (Tokyo)*, 89, 563, 1981.
175. **Mirocha, C. J., Pathre, S. V., and Robison, T. S.,** Comparative metabolism of zearalenone and transmission into bovine milk, *Food Cosmet. Toxicol.*, 19, 25, 1981.
176. **Tashiro, F., Shibata, A., Nishimura, N., and Ueno, Y.,** Zearalenone reductase from rat liver, *J. Biochem. (Tokyo)*, 11, 37, 1983.
177. **Ueno, Y., Tashiro, F., and Kobayashi, T.,** Species differences in zearalenon-reductase activity, *Food Chem. Toxicol.*, 21, 167, 1983.
178. **Yates, S. G.,** Toxicity of tall fescue forage: a review, *Econ. Bot.*, 16, 295, 1962.
179. **Jacobson, D. R., Miller, W. M., Seath, D. M., Yates, S. G., Tookey, H. L., and Wolff, I. A.,** Nature of fescue toxicity and progress toward identification of the toxic entity, *J. Dairy Sci.*, 46, 416, 1963.
180. **Keyl, A. C., Lewis, J. C., Ellis, J. J., Yates, S. J., and Tookey, H. L.,** Toxic fungi isolated from tall fescue, *Mycopathol. Mycol. Appl.*, 31, 327, 1969.
181. **Yates, S. G., Tookey, H. L., Ellis, J. J., and Burkhardt, H. J.,** Toxic butenolide produced by *Fusarium nivale* (Fires) Cesati isolated from tall fescue (*Festuca arandinacea* Schreb.), *Tetrahedron Lett.*, 7, 621, 1967.
182. **Yates, S. G., Tookey, H. L., Ellis, J. J., and Burkhardt, H. J.,** Mycotoxins produced by *Fusarium nivale* isolated from tall fescue (*Festuca arandinacea* Schreb.), *Phytochemistry*, 7, 139, 1969.
183. **White, E. P.,** Isolation of (±)-2-acetamide-2,5-dihydro-5-oxofuran from *Fusarium equiseti*, *J. Chem. Soc. (C)*, 346, 1967.
184. **Vesonder, R. F., Tjarks, L. W., Ciegler, A., Spenser, G. F., and Wallen, L. L.,** 4-Acetoamide-2-butenoic acid from *Fusarum graminearum*, *Phytochemistry*, 16, 1296, 1977.
185. **Grove, M. D., Yates, S. G., Tallent, W. H., Ellis, J. J., Wolf, I. A., Kosuri, N. R., and Nichols, R. E.,** Mycotoxins produced by *Fusarium tricinctum* as possible cause of cattle disease, *Agric. Food Chem.*, 18, 734, 1970.
186. **Ueno, Y., Sato, N., Ishii, K., Sakai, K., and Enomoto, M.,** Toxicological approaches to the metabolites of *Fusasira*. V. Neosolaniol, T-2 toxin and butenolide, toxic metabolites of *Fusarium sporotrichioides* NRRL 3510 and *Fusarium Poae* 3287, *Jpn. J. Exp. Med.*, 42, 461, 1972.
187. **Yates, S. G.,** Toxin producing fungi from fescue pasture, in *Microbiol Toxins*, Vol. 7, Kadis, S., Ciegler, A., and Ajl, S. J., Eds., Academic Press, New York, 1971, 191.
188. **Ueno, Y.,** Akakabi toxins (*Fusarium* toxins), *Jpn. J. Food Hyg. Soc.*, 14 (Part 2), 501, 1972.
189. **Tookey, H. L., Yates, S. G., Ellis, J. J., Grove, M. D., and Nichols, R. E.,** Toxic effects of a butenolide mycotoxin and of *Fusarium tricinctum* culture in cattle, *J. Am. Vet. Med. Assoc.*, 160, 1522, 1972.
190. **Bitay, Z., Glavits, R., and Sellyey, G.,** Effect of ochrotoxin A, patulin, T-2 toxin and butenolide in broiler chicken, *Acta Microbiol. Acad. Sci. Hung.*, 25, 152, 1978.
191. **Burmeister, H. R., Grove, M. D., and Knolek, W. F.,** Moniliformin and butenolide: effect on mice of high-level, long-term oral intake, *Appl. Environ. Microbiol.*, 40, 1142, 1980.
192. **Cole, R. J., Kirksey, J. W., Cutler, H. G., Doupnik, B. J., and Peckham, J. C.,** Toxin from *Fusarium moniliforme:* effects on plants and animals, *Science*, 179, 1324, 1973.
193. **Steyn, P. S. and Vleggaar, R.,** Moniliformin, a mycotoxin from *Fusarium fusarioides*, *J. Agric. Food Chem.*, 26, 375, 1978.
194. **Kriek, N. P. L., Marasas, E. F. O., Steyn, P. S., van Rensburg, S. J., and Steyn, M.,** Toxicity of a moniliformin-producing strain of *Fusarium moniliforme* var. *subglutinans* isolated from maize, *Food Cosmet. Toxicol.*, 15, 579, 1977.
195. **Springer, J. P., Clardy, J., Cole, R. J., Kirksey, J. W., Hill, R. K., Carlson, R. M., and Ishidor, J. L.,** Structure and synthesis of moniliformin, a novel cyclobutane microbial toxin, *J. Am. Chem. Soc.*, 96, 2267, 1974.
196. **Burmeister, H. R., Ciegler, A., and Vesonder, R. F.,** Moniliformin, a metabolite of *Fusarium moniliforme* NRRL 6322: purification and toxicity, *Appl. Environ. Microbiol.*, 37, 11, 1979.
197. **Thiel, P. G.,** A molecular mechanism for the toxic action of moniliformin, a mycotoxin produced by *Fusarium moniliforme*, *Biochem. Pharmacol.*, 27, 483, 1978.
198. **Marasas, W. F. O., Kellerman, T. S., Pienaar, J. G., and Nause, T. W.,** Leukoencephalomalacia: a mycotoxicosis of equine caused by *Fusarium moniliforme* Sheldon, *Onderstepoort. J. Vet. Res.*, 43, 113, 1976.
199. **Marasas, W. F. O., van Rensburg, S. J., and Mirocha, C. J.,** Incidence of *Fusarium* species and the mycotoxin, deoxynivalenol and zearalenone, in corn produced in esophageal cancer area in Transkei, *J. Agric. Food Chem.*, 27, 1108, 1979.
200. **Lin, P. and Tang, W.,** Zur Epidemiologie und Atiologie des Oesophaguscarcinoma in China, *J. Cancer Res. Clin Oncol.*, 96, 121, 1980.

201. **Gelderblom, W. C. A., Thiel, P. G., van der Merve, K. J., Marasas, W. F. O., and Spies, H. S. C.,** A mutagen produced by *Fusarium moniliforme, Toxicon,* 21, 467, 1983.
202. **Wiebe, L. A. and Bjeldanes, L. F.,** Fusarin C, a mutagen from *Fusarium moniliforme* grown on corn, *J. Food Sci.,* 46, 1424, 1981.

Chapter 3

GENETICS OF ANTIBIOTIC PRODUCTION

Zia U. Ahmed and Leo C. Vining

TABLE OF CONTENTS

I. INTRODUCTION

Antibiotics have been a major focus of interest during the past 4 decades. The most intensive studies have been aimed at exploiting the selective toxicity of these substances with the goal of developing useful new therapeutic agents to combat infectious diseases. A substantial effort has also been made to understand the mechanisms of antibiotic action, and progress in this direction has given us a satisfactory picture of the manner in which most of the important compounds exert their activity. To a degree, rapid advances in this area can be attributed to the parallel expansion in our knowledge of cell function at the molecular level and to the availability of powerful techniques for biochemical investigation.

In contrast, our understanding of the formation of antibiotics has developed at a slower pace. Only in the domain of the chemist, whose interest centers on the sequence of biosynthetic reactions that lead to the assembly of a unique natural product, have the primary objectives been achieved. Biosynthetic studies have given us a comprehensive overview of the pathways by which antibiotics are made. Here too, access to effective techniques for tracing metabolic pathways with radioisotopes, and latterly with isotopes conferring a nuclear spin that can be detected nondestructively by nuclear magnetic resonance (NMR) spectroscopy, has been a key factor. Knowledge of the enzymic processes that direct the flow of intermediates along pathways leading to antibiotic synthesis has accumulated much more slowly. Disinterest on the part of biochemists in the enzymology of antibiotic biosynthesis seems to stem from difficulties in working with these systems rather than to a lack of suitable methodology. Enzymes in secondary metabolic pathways are frequently short-lived in vivo, and this characteristic persists in vitro. Relatively little has yet been done to explain this behavior or to establish suitable procedures for isolating and purifying the enzymes of antibiotic biosynthesis.

Widespread evidence that antibiotic biosynthesis is responsive to environmental factors, such as nutrient composition and concentration, implies that the process is under metabolic control. However, our knowledge of such control mechanism is rudimentary. Constrained by the present meager base in enzymology, studies of regulatory systems have been largely oriented towards a descriptive analysis of the response of target systems. Such studies have identified control sites in a few pathways and analyzed the general nature of the regulatory processes that determine the timing and level of antibiotic production. However, the precise mechanisms by which control is effected are still unknown. Relatively little attention has yet been given to this problem.

With recent developments in gene manipulation, considerable interest has been generated in the potential application of these techniques to improving antibiotic fermentation processes and to the biogenetic construction of new active products. Fruitful use of genetic engineering procedures does not depend on prior mastery of the genetics of unknown systems, but is considerably aided where at least a basic knowledge of the system exists. Studies on the genetics of antibiotic production have accelerated in response to this need. Much work remains before the genetic systems of the principal antibiotic-producing microorganisms are adequately known. We have, as yet, only a fragmentary knowledge of the location of genes responsible for antibiotic production. The mechanisms by which gene expression is regulated in any of the antibiotic producers is still obscure. Nevertheless, rapid progress is being made. In the present chapter we outline the general aspects of antibiotic production relevant to genetic studies and survey developments in the field. The subject has been reviewed previously by Hopwood and Merrick[139] and Hopwood.[135,137] A comprehensive account of progress up to 1969 has been published by Sermonti.[323] Specific aspects of the topic have been dealt with in a large number of review articles which will be referred to in appropriate sections of this chapter.

II. ANTIBIOTIC-PRODUCING ORGANISMS

A. Diversity

By arbitrary definition,[373] an antibiotic is a low molecular weight substance produced by a microorganism and is toxic at low concentrations to other microorganisms. Even when thus restricted, the range of substances that meet these criteria is large. Current tabulations include approximately 5000 distinct compounds,[27] and the number increases year by year as the search for clinically useful agents continues. Most of these antibiotics are produced by species of bacteria and fungi. Although substances toxic to microorganisms have been isolated from unicellular algae, none has found commercial application and autotrophs, as a whole, show little promise as producers of antibiotics.

In bacteria and fungi, the capacity for antibiotic production is not uniformly distributed. Actinomycetes form the most prolific group and represent approximately 75% of the known compounds. A high proportion of the actinomycete antibiotics has been found in members of the genus *Streptomyces,* and they represent a remarkably diverse array of structural types.

The unicellular bacteria produce a more restricted range of antibiotics, with those of oligopeptide structure being the most common. Bacilli make over 200 peptide antibiotics and are the dominant producers of this type of compound.[173] In smaller numbers, antibiotics have been isolated from eubacterial families such as the Micrococcaceae, Pseudomonadaceae, Lactobacteriaceae, and Enterobacteriaceae; these compounds are more diverse in chemical structure than those from bacilli. Bacterial genera not presently included as antibiotic producers will probably be added to the list as screening for commercially useful agents is extended. The ability of many of them to make such compounds is indicated by discoveries of new antibiotics in myxobacteria:[279] *gluconobacter, chromobacterium, agrobacterium, erwinia,* and other species.[350] To some extent, the uneven distribution of antibiotic production among bacteria may be due to selective sampling during screening programs. Nevertheless, the existing observations probably reflect a more intensive development of secondary metabolic pathways in certain groups than in others.

Similar conclusions can be drawn from the manner in which antibiotic production is distributed among the fungi. From a commercial standpoint, the family Moniliaceae is of outstanding importance because of the presence within it of β-lactam producers. The class of imperfect fungi to which these species belong also produces a majority of the more than 1000 known antibiotics of fungal origin. Although the ascomycetes and basidiomycetes contain producing genera, the numbers are smaller and, notably in the basidiomycetes, the range of structural types is more limited. Phycomycetes, and also the variety of unicellular fungi classified as yeasts, have only rarely been reported to produce substances that antagonize the growth of other organisms.

B. Distribution

The majority of known antibiotic-producing organisms are soil isolates. Unquestionably, this distribution reflects the fact that actinomycetes, bacilli, and microfungi, which are the most prolific antibiotic producers, are primarily soil microorganisms. Once again, the numbers will be biased by the concentration of effort directed towards screening soil microflora for therapeutically useful substances. However, especially in recent years, when the success rate for discovering useful new drugs by traditional screening procedures has declined, a very wide range of microbial habitats has been sampled. Although antibiotic producers have been found in feces and sewage,[346] marine mud,[273] milk,[236] plant pathogens,[123] and many other substrates,[350] soil organisms remain the single most promising source of new antibiotics reported in the literature. It is difficult to avoid the conclusion that a high level of competition between the large variety of organisms that can exist in soil has selected for the evolution of a capacity to produce and excrete toxic compounds that aid in survival.

III. BIOSYNTHETIC ORIGIN OF ANTIBIOTICS

The diverse array of known antibiotics defies classification into simple structural categories. Studies on their mode of action indicate that antibiotics depend for their activity upon structural features that precisely match the shape and functional groupings at a sensitive locus in the target organism. However, the biochemical structure group to which the antibiotic belongs may be quite unrelated to the biochemical system with which it interferes.

In considering the genetics of antibiotic production, the most appropriate manner in which to classify antibiotics is by their metabolic derivation. The subtle and often cryptic link between biological activity and chemical structure means that active compounds can arise by a wide range of biosynthetic pathways. Indeed, the biosynthetic origin of antibiotics spans the entire metabolic spectrum. Any attempt at categorizing these metabolites into a few biosynthetic groups can succeed only by resorting to gross oversimplification. This will be apparent in the following subdivision into five classes based on similarities between the biosynthetic pathway by which the antibiotic is formed and those used to synthesize the major cell constituents. In general, the antibiotics discussed are those about which some information on the genetics of production is available and to which reference is made in later sections.

A. Saccharides

Antibiotics that fall logically into this class vary from simple sugar derivatives, such as nojirimycin,[167] to more complex oligosaccharides. The compounds that have attracted the most attention from geneticists are the aminocyclitols, among which are streptomycin, kanamycin, gentamycin, spectinomycin, butirosin, and others of clinical importance. The total number of aminocyclitols isolated is considerable and they occur in such different bacterial genera as *Streptomyces, Bacillus, and Pseudomonas.*[67]

The aminocyclitols also represent the only type of saccharide antibiotic on which extensive biosynthetic studies have been carried out. Pathways used in their formation have been reviewed.[283,284] Characteristic features are the derivation of components in the oligomeric structure from glucose by relatively direct conversion sequences and the linking of components by glycosidic bonds.

B. Peptides

Included in this class are numerous unusual or modified amino acids as well as modified oligopeptides and a variety of polypeptides in various size ranges. Frequently, the peptide carries a substituent, such as a fatty acid or carbohydrate, of different biosynthetic origin. The diversity of structures within the class has been cataloged and discussed;[197,366] several reviews dealing with the biosynthesis of peptide antibiotics are available.[185,191,368] The precursors may be normal protein amino acids, or they may be products diverted from a primary pathway leading to amino acids. Examples of the latter are most abundant in the shikimic acid pathway, which gives rise to antibiotics such as chloramphenicol, bacilysin, and compounds with phenazine[53] and *meta*-aminobenzyl[295] structures.

Of the peptide antibiotics, the most important are unquestionably the β-lactam derivatives. This group includes not only the classical penicillins (penams) and cephalosporins (cephems) of fungal origin, but the 7-methoxycephems, clavulanic acid (a clavam notable for its inhibition of β-lactamases), nocardicins, and carbapenems produced mainly by actinomycetes.[281] In addition, a diverse group of monocyclic β-lactams (monobactams) has now been discovered in eubacterial genera such as *Pseudomonas, Agrobacterium,* and *Gluconobacter.*[350] Penicillins and cephalosporins originate with the assembly of a linear tripeptide from α-aminoadipic acid, cysteine, and valine.[13] The catalytic mechanism is still poorly understood but may be similar to that used in the formation of glutathione.[196] The linear tripeptide is cyclized to the penam nucleus and subsequently modified to give a family of penam and cephem derivatives.

Both the clavams and the carbapenems claim glutamic acid as a precursor, although the way in which this amino acid is condensed with other precursor components indicates that the biosynthetic pathways are distinctly different.[7,8] Serine is the amino acid from which all monobactam antibiotics so far examined are derived.[157,282] At present, information on the biosynthesis of these compounds is limited to isotopic labeling data; there are no published reports on the genetics of production.

A large number of antibiotics resemble conventional oligopeptides in structure but contain unusual amino acids, especially *N*-methylated or D-forms; others contain heterologous components such as hydroxy acids or fatty acids, or severely modified forms of an amino acid (e.g., 3-hydroxypicolinic acid). Cyoligopeptide structures are also common, with cyclization being effected by either peptide or ester bond formation. Cyclization of dipeptides gives rise to a wide variety of piperazine and pyrazine derivatives which have often undergone extensive post-cyclization processing.[368]

The biosynthesis of most peptide antibiotics is not directed by ribosomes but is catalyzed by a protein thiotemplate complex. Comprehensive reviews of this process have been published.[93,183-185,194,203] Noteworthy from the viewpoint of genetic analysis is the participation of very large multifunctional enzyme proteins constructed from single polypeptide chains. Numerous catalytic domains in the tertiary structure are associated with specific enzyme activities. Little genetic research has been carried out with peptide-antibiotic producers despite the fact that *Bacillus subtilis,* a member of the dominant producing genus, offers a convenient model for genetic analysis.

Although most of the peptide antibiotics investigated appear to be biosynthesized on protein templates, exceptions have been noted. Nisin in *Streptococcus lactis*[160] and subtilin in *B. subtilis*[252] are probably made by excision from ribosomally synthesized polypeptides that are modified after translation to generate abnormal amino acid components (e.g., lanthionine). No biosynthetic or genetic studies have yet been carried out on the considerable number of large polypeptide and protein antibiotics that have been described,[366] but these are very likely made under ribosomal direction as well.

C. Acetogenins

A remarkable diversity of biochemical structures can be created by the self-condensation of activated acyl units. The archetypal precursor is acetyl-CoA and two main subgroups of antibiotic acetogenins can be recognized: those derived directly by linear head-to-tail condensation of acyl units (comparable to the fatty acids of primary metabolism) and those derived indirectly via an active isoprenoid intermediate. The isoprenoid pathway yields a family of terpenoid and steroid products in both primary and secondary metabolism. Some of the secondary metabolites (e.g., fusidic acid[108]) have antibiotic activity; fungi are the dominant producers of such compounds. The genetics of their biosynthesis has not yet been investigated.

Of the antibiotics formed via head-to-tail acyl condensations, the group with linear alkyl chains, such as cerulenin[278] and the polyacetylenes,[50] bear a close biosynthetic relationship to the fatty acids. Although it is clear that the primary and secondary metabolites are formed by similar processes, with acetyl-CoA or an analog acting as the chain initiator and malonyl-CoA as the chain-extending intermediate, details of the secondary biosynthetic mechanism, particularly at the enzyme level, are meager.

Acetogenins in which a linear polyketide intermediate has cyclized to generate an aromatic structure are relatively common among both actinomycete and fungal secondary metabolites. Examination of the enzymic mechanism by which 6-methylsalicylic acid is made in *Penicillium griseofulvum* has implicated a multifunctional protein with similarities to that used for polypeptide synthesis.[72,202] Included among the aromatic acetogenins are the clinically important tetracyclines, which are produced by several species of *Streptomyces;*[158] their biosynthesis has been established by a wealth of biochemical and genetic data.[26,161,237] The anthracycline family of anthraquinone glycosides are also aromatic acetogenins produced by streptomycetes. The biosynthesis of these important anti-cancer agents has been reviewed.[161,381] Among the aromatic acetogenins of fungal origin, griseofulvin, which is produced by *P. griseofulvum,*[144] has been used to treat persistent mycotic infections.

A large group of acetogenin macrolides differ from the preceding compounds in that the polyketide intermediate is formed predominantly from methylmalonyl chain-extending units.[274,275] Instead of cyclizing to an aromatic structure, this intermediate is partially reduced, generating hydroxylic functions. One such hydroxyl becomes esterified with the terminal carboxyl group; others are glycosidated to give such commercially important macrocyclic lactone antibiotics as erythromycin (in *Streptomyces erythreus* and *Arthrobacter* sp. NRRLB3381) and tylosin (in *S. fradiae*).

One important subgroup of the macrolides differs principally in containing a sequence of up to seven conjugated double bonds within the macrocyclic ring structure and in showing marked antifungal rather than antibacterial activity. The biosynthesis of these "polyenes" follows a pattern similar to that of the macrolides but a higher proportion of adjacent two-carbon acyl precursors is incorporated into the polyketide intermediate.[229] Candicidin, a product of *S. griseus,* is the most intensively investigated member of this group.

A second subgroup of macrolides, the ansamycins, can be distinguished by the presence within the structure of an aromatic moiety derived from a *meta*-aminobenzyl chain-initiating unit.[195] Variations in ansamycin structures arise partly from differences in the number and type of acyl chain-extending units involved in their biosynthesis. Of prime interest are the rifamycins which, in chemically modified form, are of considerable value as therapeutic agents.[103]

The polyethers form a separate large group of actinomycete acetogenins bearing a close biosynthetic relationship to the macrolides. Although they are formed, like the macrolides, by condensation of mixed 2-, 3-, and 4-carbon acyl precursors followed by partial reduction of the polyketide intermediate, subsequent ring-forming reactions generate ionophoric cyclic ether systems instead of lactones.[379,380] Monensin and other polyether antibiotics have found important applications as coccidiostats in the poultry industry and as feed additives with growth-promoting activity. In contrast, a related group of polyketide-derived ionophoric antibiotics, the macrotetrolides, has attracted less interest.[25]

D. Nucleosides

Within this class of antibiotics, several subgroups can be recognized.[349] In compounds such as the blastmycins, polyoxins, and psicofuranine, a normal purine or pyrimidine base is attached as an *N*-glycoside to an unusual sugar moiety, or one carrying unusual substituents. Other nucleoside antibiotics have an abnormal heterocyclic base in addition to unusual sugar components and additional substituents. The biosynthesis of some compounds of this type has been examined through isotopic labeling. In the case of pyrrolopyrimidines such as sangivamycin, the primary precursor is a normal nucleoside which is modified to generate the abnormal structure. The group of carbon-linked nucleoside antibiotics, such as formycin and showdomycin, appear to be biosynthesized *de novo* from α-ketoglutaric acid. The amount of published information on the biosynthesis of nucleoside antibiotics and its genetic determinants is disappointingly small.

E. Heteromers

This is a miscellaneous group, included to provide a home for compounds such as the tetramic acids and antimycins, which are derived from the mixed condensation of amino acid and acyl precursors.[100,249] It can also accommodate the novobiocin-coumermycin group,[188] glycopeptides of the vancomycin-ristocetin type,[47] and numerous other antibiotics of heterogeneous biosynthetic origin, such as prodigiosin.[102]

IV. CHARACTERISTICS OF PRODUCTION

The main thrust in studies on the genetics of antibiotic production has been to improve fermentation yields. One route to success lies in manipulation of metabolic control systems; a fuller understanding of the mechanisms that regulate antibiotic production is, therefore, an important component of the genetic approach. The following section briefly summarizes the characteristics of antibiotic production.

A. Trophophase and Idiophase

Antibiotics are not usually produced in a fixed ratio to growth. Production varies with the kind of medium in which the organism is grown and often also with the stage of growth. It is widely observed that the formation of antibiotics and other secondary metabolites in batch culture takes place after most of the biomass has been produced, giving rise to a biphasic fermentation pattern. The terms "trophophase" and "idiophase" identify the growth and product-accumulation stages, respectively.[51] In general, secondary metabolites are formed only when growth is restricted; a medium formulated to support balanced growth until near saturation and then to exhaust a selected component imposes a period of unbalanced growth at high cell density on the organism. The shift down favors antibiotic production and the pattern of such fermentations is clearly biphasic. However, culture media with one of the essential nutrients in a form that can be only slowly assimilated create unbalanced growth conditions from the outset. Secondary metabolites are then produced during slow accumulation of biomass, giving a uniphasic fermentation pattern. In complex media with multiple nutrient sources, the pattern is usually a composite of these two extremes.

B. Catabolite Repression

While they are available in the medium, readily assimilated sources of carbon and nitrogen promote cell growth and usually have the reverse effect on antibiotic production. In many microorganisms, nutrients preferred for growth suppress the assimilation of alternative carbon and nitrogen sources. The term "catabolite repression" used to describe this latter phenomenon is commonly applied as well to the suppression of secondary metabolism by preferred growth substrates.[99] Whether the same regulatory mechanism produces both effects has not been determined, but it is a plausible possibility. Rapid growth is promoted by suppressing superfluous metabolically expensive activity, such as the synthesis of antibiotics that are not needed for protection when the organism is growing rapidly.[365] An alternative suggestion[52] is that secondary metabolism, along with differentiation processes such as sporulation, is encoded in a "nonvegetative genome", which is controlled separately from the genes needed for primary metabolism. This nonvegetative genome would be negatively regulated by products of vegetative growth. However, the mechanisms of catabolite repression in either of its intended contexts are poorly understood in antibiotic-producing organisms. Genetic analysis, which will be essential to a thorough understanding of the phenomenon, has not yet been attempted.

C. Phosphate Regulation

Limiting the amount of phosphate in the nutrient medium is frequently an essential requirement for optimum antibiotic production.[228] Enzyme synthesis is believed to be the main target of the phosphate effect, but inhibition of enzyme activity may also be a factor, at least in candicidin production.[231] Where phosphorylated intermediates are known to be involved, as in the pathway for streptomycin and neomycin biosynthesis, suppression of phosphatase activity has been suggested as a possible control mechanism.[220] However, antibiotics such as tetracyclines, macrolides, and β-lactams not known to have phosphorylated biosynthetic intermediates are also subject to phosphate regulation.

There is evidence that adenosine triphosphate (ATP) concentration in some organisms is inversely correlated with antibiotic production, and this nucleotide has been postulated to function as an intracellular effector for secondary metabolic processes.[204] Some progress has been made in identifying antibiotic pathway enzymes specifically repressed by phosphate,[71] and the isolation of phosphate-deregulated mutants promises to bring a better understanding of the mechanism of phosphate control.[233]

D. Growth Rate Control

Some antibiotics are produced in response to a decreased growth rate imposed nonspecifically by limiting the carbon, nitrogen, or phosphate source.[201] In these fermentations, control of antibiotic biosynthesis is presumed to be inversely linked to growth rate per se. The control mechanism is not known.

E. Induction

Methionine has been reported to induce the production of cephalosporin in *Acremonium chrysogenum*[79] and there is unequivocal evidence for the inudction of some enzymes in the pathway of patulin biosynthesis.[101] However, 6-methylsalicylic acid synthetase, the initial enzyme in the patulin pathway, is not induced and, in general, secondary pathway enzymes do not seem to be induced by abnormal accumulations of primary precursors, although processing enzymes that modify the initially formed secondary metabolite may be inducible.

F. Bioregulators

A separate phenomenon referred to as ''induction'' is mediated by a bioregulator unrelated in structure to pathway intermediates.[182] The existence of such bioregulators was first reported in *S. griseus,* where certain mutants lost both the capacity to produce streptomycin and to sporulate. The normal phenotype was restored by trace amounts of A-factor, a substance excreted by wild-type strains.[176] A-factor has been characterized as 2(*S*)-isocapryloyl-3(*R*)-hydroxymethyl-γ-butyrolactone. It is produced by various species of *Streptomyces* and is also essential for production of the anthracycline-type antibiotic, leukemomycin, by *S. griseus.*[110]

The mechanism of action is not well understood. Reports that A-factor intervenes in carbohydrate metabolism to augment the supply of glucose-6-phosphate for streptomycin biosynthesis have not been substantiated.[118] A-factor induces resistance to streptomycin as well as production of the antibiotic in *S. griseus* and is required for sporulation in this species. However, in *S. lividans*[153] and in *S. coelicolor* A3(2), which also produce A-factor, the bioregulator is required neither for sporulation nor for synthesis of undecylprodigiosin and the polyketide antibiotic actinorhodin, which are characteristic products of these organisms.[119] Interestingly, however, the *S. coelicolor* A3(2) chromosome appears to contain a pleiotropic regulatory gene (*afsB*) that positively controls the formation of A-factor and the two antibiotics.[153]

G. Precursor Control

Antibiotics compete with normal cell metabolic processes for their supply of precursors. This competition has been well documented for actinomycin where protein biosynthesis competes for certain amino acids,[174] and may account in part for the incompatibility between fast growth of a culture and optimum output of an antibiotic. Moreover, the close feedback control normally exerted over the biosynthesis of cell macromolecule precursors means that antibiotic pathways sharing such precursors are influenced by primary control systems.[109]

In *Penicillium chrysogenum,* penicillin production is low in media containing lysine. The amino acid suppresses its own biosynthesis by feedback inhibition, thereby preventing formation of α-aminoadipic acid, which is both an intermediate in the lysine pathway and a precursor for penicillin biosynthesis.[90,340] In the actinomycetes, where α-aminoadipic acid is a product of lysine metabolism instead of an intermediate in its biosynthesis,[382] excess lysine is beneficial to β-lactam antibiotic production. Among *S. clavuligerus* mutants resistant to the lysine analogue, *S*-(2-aminoethyl)cysteine, those lacking allosteric feedback control of aspartokinase activity produce up to sevenfold more cephamycin than the wild type.[239,240]

The formation of candicidin by *S. griseus* requires *p*-aminobenzoic acid, which is supplied by the branched primary pathway for aromatic amino acid biosynthesis. Because this pathway is controlled through end product repression and inhibition of the initial enzyme, deoxyarabinoheptulosonate phosphate synthetase, aromatic amino acids indirectly suppress candicidin production.[206] However, the amino acids also repress synthesis of a specific *p*-aminobenzoic acid synthetase needed to form candicidin.[230] This is a clear example of a direct rather than a competitive effect of primary metabolites on the secondary process.

H. Feedback Effects

Antibiotic synthesis in some xenotoxic systems,[365] such as the penicillin fermentation with *P. chrysogenum,* can be prolonged by slow feeding of an unbalanced nutrient supply to maintain appropriate idiophase conditions. However, the period of active antibiotic synthesis in many fermentations is relatively short. Provided the carbon source is in excess, production rarely terminates because of cell lysis; it is more likely to be limited by either enzyme decay or feedback effects from the accumulated antibiotic. The enzymes of secondary metabolic pathways are frequently unstable in vivo as well as in vitro. In one of the few studies of this phenomenon, Friebel and Demain[89] showed that inactivation of gramicidin-S biosynthetic enzymes could be prevented by anaerobiosis, implicating oxidation of the numerous thiol groups in peptide synthetases as the chief reason for loss of activity. Since the soluble peptide synthetases of bacilli become associated with membranes during the late growth phase, changes in activity may also be related to changes in cell structure as sporulation occurs.[364] In the fungus *P. griseofulvum,* the decay of enzymes specific for patulin biosynthesis is at least partly due to proteolytic action.[251] The reason for the apparently selective sensitivity of secondary pathway enzymes is not yet clear.

Feedback effects exerted by antibiotics on their own production have been observed in a number of fermentations.[69] The mechanism for such control is particularly intriguing with autotoxic antibiotics where the product is efficiently excreted and its intracellular concentration is probably too low to control the biosynthetic process directly. In examples where feedback has been examined, the mechanism appears to vary. The initial enzyme of chloramphenicol biosynthesis in *S. venezuelae* is repressed by high broth concentrations of the antibiotic,[171,367] whereas in *S. alboniger,* methylation of the penultimate intermediate in the biosynthesis of puromycin is inhibited by the end product.[308] Inhibition of aurodox biosynthesis in *S. goldiniensis* by the antibiotic and its structural analogs is reversible, and thus is more likely due also to enzyme inhibition than to repression.[205]

V. GENETIC ANALYSIS OF PRODUCERS

A. Fungi

The organisms that have come under close genetic scrutiny are those that make β-lactams. Fungi that produce griseofulvin and fusidic acid, two antibiotics of clinical value but of relatively minor commercial importance, have received only cursory attention.[54]

β-Lactams are produced in small amounts by a variety of fungi, including *Aspergillus nidulans.* The presence of a sexual stage and the well-developed procedures for genetic analysis in this ascomycete have facilitated studies on the genetics of penicillin production.[213] In *A. nidulans,* genetic recombination can occur in both the sexual and parasexual cycles.[287] Since this is a homothallic fungus, sexual structures develop from the fusion of two haploid nuclei originating in the same hyphae. Exchange of genetic material occurs if two strains are grown together, yielding a dikaryotic mycelium by hyphal anastomosis. Syngamy produces three types of diploid cells, two homozygous and one heterogygous, each of which gives rise to cleistothecia containing sets of ascospores packaged within asci. By selecting hybrid cleistothecia, the incidence of recombinant progeny in colonies arising from separated ascospores can be determined.

Neither *P. chrysogenum* nor *A. chrysogenum,* the two fungi used commercially to produce β-lactam antibiotics, has a sexual stage, but genetic analysis can be performed via the parasexual cycle.[212] In a heterokaryotic mycelium, mitotic division of heterozygous diploid nuclei generates occasional recombinant genotypes that can be recognized and analyzed. Non-disjunction of a centromere during mitosis results in unstable monosomic (2n-1) nuclei that revert to haploidy by random loss of individual chromosomes from the disomic pairs. This haploidization can be induced with agents such as *p*-fluorophenylalanine[213] and yields progeny showing free recombination between whole chromosomes of the original haploid parents. Haploidization analysis thus provides information on the presence of genes in linkage groups. Mitotic crossing over between homologous chromosomes of the heterozygous diploid gives strains that are homozygous for loci further from the centromere than the crossover point. Analysis of such segregants provides linkage data on genes relative to known markers within the chromosome.

The ability of fungi to generate heterokaryons by hyphal anastomosis also furnishes a useful test for complementation and dominance.

B. Unicellular Bacteria

The natural processes of genetic exchange in this group, conjugation, transduction, and transformation, are relatively well known. Conjugation is widely distributed but only the pseudomonads, among those genera in which it has been demonstrated, are noteworthy as antibiotic producers. Generalized transduction is also available for the genetic analysis of *Pseudomonas aeruginosa* which, although well developed,[129] has not yet included genes for secondary metabolism. Conjugation occurs in the streptococci and less commonly in the bacilli. Genetic analysis of the latter, which constitute the most prolific antibiotic-producing group of unicellular bacteria, has been developed extensively through the use of transduction and transformation; fusion of protoplasts has also been employed recently to mix genomes.[87,309] Transformation has been shown to occur in *Bacillus licheniformis, B. pumilis,* and *B. subtilis,*[341,389] which are all reported to produce antibiotics, and in the streptococci.[76] Generalized transduction has also been used for genetic analysis in *B. subtilis*[199] in which antibiotic production has received some attention as a genetic character, but even here interest has focused mainly on its relationship to endospore formation.

C. Actinomycetes

A substantial amount of genetic information has now been obtained on the genus *Streptomyces,* which contains a high proportion of the known antibiotic producers. *Nocardia, Micromonospora,* and *Mycobacterium* species have also been investigated; the last-named genus is not significant as a producer of antibiotics.

1. Conjugation

Streptomyces are able to exchange material by conjugation. The process has been most intensively studied in *S. coelicolor* A3(2), which is recognized to be a strain of *S. violaceoruber.*[141,192] The actual mechanism by which DNA is transferred between strains is not known, but a conjugative process is inferred because physical contact between vegetative hyphae is necessary and substantial regions of the genome are recombined.[141]

Transfer of complete genomes, perhaps by a process analogous to hyphal anastomosis in fungi, can occur since *Streptomyces* are able to exist as heterokaryons. Strains differ in the ease with which heterokaryotic mycelium forms. In *S. griseus* and some other species, it was the most abundant type detected when two strains with different auxotrophies were grown together under conditions selecting for prototrophy.[43,45] In *S. rimosus*[91] and *S. glaucescens,*[24] the initial prototrophic growth contained a mixutre of heterokaryons and true recombinant forms. The latter can be detected by the absence of prototrophic progeny after propagation via spores, which in *Streptomyces* are generally uninucleoidal.[44,138] In *S. coelicolor* A3(2) on nonselective media, heterokaryons are relatively uncommon[140] but develop from mixed platings of differently auxotrophic SCP1$^-$ SCP2$^-$ strains on minimal medium lightly supplemented with growth factors.[241] In strains containing either fertility plasmid, heterokaryosis is suppressed. Pairs of *S. venezuelae* auxotrophs mixed on minimal medium also form heterokaryons readily when the combination does not elicit "lethal zygosis" (see below); if it does, only recombinants appear.[390] As in fungi, the characteristics of heterokaryons, when they form, can provide useful information on dominance and complementation.

Genetic analysis of actinomycetes has, until recently, been based exclusively on conjugative mechanisms of marker transfer and enumeration of haploid recombinant classes on selective media. Mapping procedures developed with *S. coelicolor* A3(2)[132] have been extended to other streptomycetes[4,5,9,24,66,91,178,234] and to *Nocardia mediterranae;*[320] the linkage data are consistent with the presence of circular genomes in these organisms.[148] Because of the relatively large size of the DNA segments exchanged during conjugation, gene spacing cannot normally be estimated from the analysis of haploid recombinants. However, in some situations the conjugative process yields partial diploids as well as heterokaryons and recombinant haploid progeny. The partial diploids form colonies (heteroclones) consisting of a mixture of recombinant and parental genotypes[329] which are postulated to arise before sporulation by additional cross-over events in a terminally redundant merozygote formed initially by a single cross-over (or an unequal number of cross-overs) between complete and incomplete parental genomes.[133] The genotypes then segregate into spores. Analysis of genotypes within the heteroclone gives information on relative recombination frequencies between markers, enabling gene sequences to be determined and map distances to be calculated.[140]

Heteroclone analysis has been of value in the genetic mapping of *S. rimosus*[6] and *Actinomyces (Streptomyces) olivaceus,*[234] as well of *S. coelicolor* A3(2).[131,132] In accord with their postulated origin by an unequal number of cross-overs, heteroclones are generated more frequently in *S. coelicolor* A3(2) when the initial selection is for two closely linked markers.[140] However, they can be isolated from crosses lacking this feature by a procedure involving a cellophane underlay that facilitates the separation of heteroclones from other progeny.[6,326] Because of the partial heterozygosity of the initial genomic material, heteroclones can be used to test for complementation and dominance.[120,121]

2. Fertility Plasmids

The first analysis of fertility differences in *S. coelicolor* A3(2), carried out by Sermonti and Casciano,[324] suggested two fertility types. A third type was added by Hopwood et al.[149] and the three types were designated IF, NF, and UF.[370] The IF fertility type has been equated with the presence of plasmid SCP1 carried autonomously; in NF strains, SCP1 is believed to be integrated within the chromosome and in UF strains it is absent.[369] Despite intensive effort, plasmid SCP1 has not yet been isolated. Crosses between $SCP1^+$ (IF) and $SCP1^-$ strains give approximately 10^{-5} recombinants among the total number of colony forming units (cfu) harvested. Marker transfer is nonpolar whereas all $SCP1^-$ strains are converted to IF types. Thus, SCP1 is transmissible and its infectious transfer is associated with reciprocal gene exchange; either parent may serve as the recipient and contribute all of its chromosomal DNA to the merozygote. In contrast, crosses between $SCP1^-$ and $SCP1^+$ (NF) strains show polar transfer of chromosomal markers with recombination frequencies close to 100%. All progeny are $SCP1^+$ (NF), as expected if an integrated plasmid and chromosomal markers are transferred as a unit.[148]

In $SCP1^- \times SCP1^+$ (NF) crosses, a bidirectional gradient in the frequency of marker transfer is observed, with the 9 o'clock map position representing 100% and the frequency decreasing in either direction to as low as 1% for distant markers.[143] Vivian and Hopwood[371] also isolated unstable strains that donated to $SCP1^-$ recipients a unidirectional marker gradient always originating at a fixed map position. Although the progeny were all $SCP1^+$, transfer of chromosomal markers was below 100% and it was suggested that the donors contained an autonomous SCP1 plasmid that had acquired an unstable chromosomal insert; transient interaction between this insert and the homologous region in the donor chromosome mobilized and transferred DNA segments in crosses with $SCP1^-$ strains.

Donor strains that contain a recognizable chromosomal marker integrated into SCP1 have also been obtained.[142,144] Because of the parallel with F′ strains of *Escherichia coli,* these are designated SCP1′. An SCP1′-*cysB* strain transferred the plasmid to SCP1$^-$ strains infectiously but did not mobilize chromosomal markers, for which it showed fertility characteristics resembling SCP1$^+$ (IF) strains.

Much of the residual fertility in SCP1$^-$ × SCP1$^-$ crosses could be ascribed to a second plasmid in *S. coelicolor* A3(2).[32] This plasmid, designated SCP2, has been isolated and shown to have a size of 20 Mdaltons.[32,316,318] A variant form identified as SCP2*, which confers higher fertility on its host than SCP2 but has the same size and restriction map, has also been reported.[31] The remaining low level of fertility (approximately 10^{-7} recombinants per cfu in crosses with SCP1$^-$ SCP2$^-$ strains) may be due to genetic elements integrated into the *S. coelicolor* A3(2) chromosome.[137] One of these, SLP1, has been detected as an autonomous plasmid in *S. lividans* 66 after excision from a region close to *strA* in the *S. coelicolor* A3(2) chromosome during interspecific mating, and also after in vitro introduction of DNA form SCP1$^-$ SCP2$^-$ strains of *S. coelicolor* A3(2).[36] Plasmid SLP1 has been isolated and shown to occur in a range of sizes, identified as SLP1.1, SLP1.2, etc. In addition, evidence of a specific "lethal zygosis" phenotype (see next section) different from that given by SLP1 indicated that another *S. coelicolor* A3(2) replicon, SLP4, was also present in *S. lividans* 66 after interspecific matings.[150]

In *S. lividans* 66 itself, the presence of two fertility plasmids, SLP2 and SLP3, was also inferred from the manifestation of "lethal zygosis" between parental strains and certain regenerated protoplasts. The presence of two plasmids was supported by cross-fertility tests; strains identified as SLP2$^-$ SLP3$^-$ were infertile whereas SLP2$^+$ × SLP2$^-$ crosses gave 10^{-4} to 10^{-5} recombinants per cfu, but whether SLP3 also contributes to the ability of *S. lividans* 66 to recombine chromosomal markers by conjugation is uncertain; apparently, neither plasmid has been isolated.[150]

A second *S. lividans* strain, ISP5434, contains a small (5.9 Mdalton) plasmid, pIJ101, which enhances genetic exchange when transferred into *S. lividans* 66 or *S. coelicolor* A 3(2).[177] Recombination frequencies of 10^{-2} per cfu were obtained, making possible mapping procedures that do not depend on the use of selective media. The plasmid has a broad host range and was transferred into 12 of 15 other *Streptomyces* species tested. Although it is accompanied in *S. lividans* ISP5434 by three smaller plasmids, pIJ102, pIJ103, and pIJ104, with DNA sequences in common, only pIJ101 is conjugative and elicits the "lethal zygosis" reaction.

Fertility is also plasmid mediated in *S. rimosus*. Here, elimination of plasmid SRP1 markedly reduced the frequency of recombination.[92] In *S. erythreus* NRRL2338, fertility may be mediated by a plasmid associated with erythromycin biosynthesis,[376] but other interpretations of the data are possible. Putative plasmid-mediated fertility is also present in *S. venezuelae* ISP5230. Auxotrophic derivatives of the wild type are infertile, but two fertile strains, VS191 and VS192, were recovered from one auxotroph (VS113) using the "lethal zygosis" technique of Bibb et al.[36] A third fertile strain, VS194, was subsequently isolated from VS191 by the same procedure.[391] However, no plasmid has yet been detected in DNA from *S. venezuelae* ISP5230 strains.

3. Lethal Zygosis

In some streptomycetes, plasmid-containing strains inhibit the growth or sporulation (or both) of plasmidless counterparts. The phenomenon has been called "lethal zygosis" because of a similarity to the lethal effect of plasmid transfer between Hfr and F$^-$ strains in *E. coli*,[337] although use of this term is not intended to imply that the mechanisms are identical.[32] Inhibition of the plasmidless variant seems to require hyphal contact leading to conjugation between the two strains and was first observed in streptomycetes with strains of *S. coelicolor* A3(2) that differed only in the presence or absence of plasmid SCP2*.[32] Interactions of this kind have been used as a convenient operational test for the elimination of a transmissible plasmid from a plasmid-bearing strain. When they are observed between a parent strain and progeny recovered after procedures known to cause plasmid loss, they are also taken as evidence for the existence of plasmids undetectable by physical methods.

In practice, the strains are grown together so that the plasmid bearer develops in a confluent lawn of the plasmidless variant. The perimeter of the colony is marked by a narrow zone in which development of the plasmidless strain has been retarded. Where the plasmid-containing colony develops from a point source, as when protoplasts are exposed to plasmid DNA and the transformation mixture is plated on regeneration medium, the inhibited region is usually small and is referred to as a ''pock''.[35] Several uses of pock formation to detect progeny that have taken up a conjugative vector during cloning procedures have been described.[28,186,259] The formation of pocks has also been used to detect conjugative transfer to *S. lividans* 66 of plasmids isolated from other streptomycetes.[244,334] Pock morphologies are often characteristic of the plasmid and are reproduced with plasmid preparations from a new host. Only a minority of foreign plasmids will form pocks in *S. lividans* 66 and a positive result serves to identify those potentially useful as cloning vectors in this and possibly other species.

Besides *S. lividans* 66 and *S. coelicolor* A3(2), in which pock formation has been frequently reported,[35] strains of *S. lavendulae, S. parvulus, S. actuosus,* and *S. roseochromogenes* give pocks when serving as recipients for the pSRC plasmids of the latter species, as does *S. chartreusis* SF1623 during transfer of the foreign plasmids pSF601, pSF674, pSF689, and pSF765. These reports, and others of pock formation in *S. noursei* 2217-G with plasmid pSCY3,[351] in *S. espinosus* with plasmid pUC6,[224] and in *S. lavendulae* with plasmid pTA4001[186] indicate that ''lethal zygosis'' in streptomycetes is a fairly widespread phenomenon.

Ogata et al.[259] have observed spontaneous formation of pocks in confluent lawns of *S. endus* KCCS-0123. Lysis of aerial mycelium and suppression of sporulation was associated with the presence of particles resembling phage tails, and the possibility that a defective phage might be responsible for the phenomenon was suggested. Similar particles were found, along with intact phage, in plaques that developed spontaneously in *S. azureus* ATCC14921 during subculture. In the latter, the presence of a temperate phage, SAt2, which generated defective as well as virulent complete structures, was considered to be responsible.[260] The possible association of incomplete phage gene products with ''lethal zygosis'' has been considered in *S. fradiae* where a large (24-kilobase) deletion in plasmid pUC13, the 76-kilobase prophage form of 0SF1, was associated with a pock-like interaction not seen with the intact plasmid.[63,64]

4. Protoplast Fusion

In microorganisms lacking convenient natural systems for bringing genetic material from two strains together in the same cytoplasm, the barrier to recombination can often be overcome by preparing protoplasts and promoting fusion before the cellular form is regenerated.[18,152,257] Fusion is promoted by adding polyethylene glycol. When a fusion mixture of protoplasts from genetically marked strains of *S. coelicolor* A3(2) was allowed to sporulate on regeneration medium, Hopwood et al.[152] detected haploid recombinants at frequencies above 20% in the spore population. Recombinants were also obtained from *S. lividans, S. griseus,* and *S. parvulus,* whereas the procedure was less successful with *S. acrimycini.* Conditions for protoplast fusion have been developed independently by Baltz[18,22] and give recombination frequencies of up to 15% of the total viable protoplasts with *S. fradiae* and *S. griseofuscus*. This procedure also works well for other streptomycetes and for a *Streptosporangium* species.[19] Related fusion protocols have since been used to obtain recombinants from a number of actinomycetes,[246,258] and even higher recombination frequencies have been achieved by irradiating the protoplasts with ultraviolet light.[19,146]

Linkage analysis after fusion of *S. coelicolor* A3(2) strains carrying markers well spaced around the chromosome gave recombinant types indicating that complete genomes of each parent had participated.[145] The results also showed that analysis of recombinants is complicated by opportunities for repeated crossing over while the diploid condition persists. The effect of successive rounds of recombination is to cause an apparent increase in map distance so that markers separated by only 10 to 15% of the genome may appear unlinked.[147] Although the procedure is therefore of limited use for determining marker sequence and map intervals,[137] it has been used successfully with genetically marked strains of *S. fradiae* to place 11 markers on a genetic map of the chromosome.[19,21]

Fusion of protoplasts is of considerable value for preparing strains with existing genetic markers assembled in new combinations. Because of the high rate of recombination, progeny from a cross can be isolated on nonselective media, making it possible to recover strains bearing all or a selected group of the markers present in both parents. Protoplast fusion is expected to be an effecive means of reassembling in new combinations the genes responsible for overproduction of antibiotics,[193] and its potential for bringing the DNA of two cells together within one cytoplasm makes it possible to examine the feasibility of producing hybrid species (and thus hybrid antibiotics) by interspecific recombination. This has been explored with *S. lipmanii* and *S. jumonjinensis* (see Section VI.A) and also with *S. fradiae* and *S. bikiniensis*. In the latter, protoplast fusion raised the recombination frequency above that obtained in conventional crosses (10^{-7} to 10^{-8} per cfu) by 200- to 10,000-fold, depending upon the marker examined.[107] The prospect of developing new antibiotics with hybrid structures by this technique can thus be explored.[137]

4. *Transduction*

An early report of transduction in *S. olivaceus*[10] has not been extended and several later studies with actinophages gave no indication of chromosomal gene transfer.[49,78,170,209,266,388] However, an effective system has been discovered for *S. venezuelae* where the temperate phage SV1[343] carries out generalized transduction.[342] Moreover, phage grown on the wild-type strain, mutagenized in vitro with hydroxylamine, and used to transduce a *lysA* mutant to *lysA*$^+$, introduce mutations for adjacent *met, thr,* and *hom* genes associated with biosynthesis of the aspartate family of amino acids.[344] Similar experiments have revealed close linkage between *his* and *trp* loci.[345] In showing the cotransducibility of neighboring genes in *S. venezuelae,* these experiments establish the feasibility of fine-structure genetic mapping of actinomycete genomes by such techniques.

Generalized transduction has also been reported recently in a neomycin-producing strain of *S. fradiae* using the temperate actinophage SF1, which resembles *E. coli* phage P1 in being maintained within the cytoplasm as a plasmid prophage.[64]

5. *Transformation*

Preliminary evidence[37] for the transformation of intact mycelium of non-producing streptomycetes to antibiotic biosynthesis using DNA from producing strains has not been confirmed and *Thermoactinomyces vulgaris,* in which transformation of genetic markers has been demonstrated, is now classified as a bacillus.[141] Only one other report of DNA uptake into intact actinomycete mycelium has appeared; in this, Xue et al.[386] have described the transformation of calcium-treated *S. griseus* cells with plasmid DNA (see Section VI.B.3). Transformation (transfection) of streptomycete protoplasts with plasmid or phage DNA in the presence of polyethylene glycol is discussed later (see Section VII.D).

VI. GENETIC ALTERATION OF STRAINS

A. Natural Instability

Streptomycetes exhibit a high degree of instability in antibiotic production, as well as in other traits. Although this has been widely recognized, past efforts have been directed more towards overcoming the attendant problems than towards understanding its causes. Recently, however, the characteristics of genetic variability have been defined in a variety of organisms and progress has been made towards determining the molecular mechanisms.

The existence of a genetic system in *S. coelicolor* A3(2) conveying instability was indicated in early studies of fertility,[143,371] and was demonstrated more explicitly in strains resistant to chloramphenicol (*Cml*^{R}), which gave rise to sensitive variants (*Cml*^{S}) at frequencies of 0.5 to 2.0%, much higher than could be accounted for by mutation. Since the sensitive variants reverted (although at a much reduced frequency) to resistance, which was again lost at the same high rate, Freeman et al.[88] concluded that the conversion was due to a rearrangement rather than to a permanent loss of genetic material. The trait responsible for the reversible phenotypic behavior could not be mapped to a unique position on the chromosome, nor could it be located on the recognized plasmids SCP1 and SCP2. It was suggested that a mobile gene, perhaps capable of transient extrachromosomal existence, might be involved. Similar results were obtained independently by Sermonti et al.,[330] who subsequently investigated the genetics in more detail.[327,328,331] Although Cml^R was present at single unequivocal sites in fresh, single-colony isolates and reverted Cml^S strains, the map position varied. The resistant determinant could also reversibly transfer to SCP1, where it became associated with fertility, and was usually linked to genes controlling resistance to tetracycline and rifamycin, aerial mycelium formation, and synthesis of arginine. In some strains, arginine auxotrophy, caused by loss of *argG*-coded arginosuccinate synthetase activity, became dissociated from the Cml^R trait. A transposable element, SCTn1, was postulated to account for these characteristics; however, the recombination mechanism was not determined.

The observation that arginine-requiring mutants appear at unexpectedly high frequencies in strains exhibiting genetic instability has been reported with other actinomycetes. Cultures of *S. scabies, S. alboniger,* and *S. violaceoruber* treated with intercalating dyes yielded a high proportion of strains lacking aerial mycelium (Amy$^-$) and many of these were also arginine auxotrophs due to a concurrent loss of arginosuccinate synthetase activity.[291,292] Althought the Amy$^-$ Arg$^-$ strains never reverted, it was considered unlikely that a primary biosynthetic enzyme would be plasmid-borne and the phenotype was postulated to have been caused by integration of a DNA element within a region of the chromosome specifying arginosuccinate synthetase. The possibility that the DNA insert originated on a plasmid was suggested but no plasmid with the predicted characteristics has been isolated.

A similar type of insertion mechanism has been proposed to explain the variability of β-lactamase production in streptomycetes.[261] In *S. lavendulae,* the high-frequency loss of β-lactamase production caused by treatment with acriflavine[262] can sometimes be accompanied by a reversible loss of arginosuccinate synthetase activity and inability to sporulate. Matsubara-Nakano et al.[235] suggested insertional inactivation of the *argB* gene, possibly with DNA coding for β-lactamase forming the mobile element. Involvement of a plasmid was discounted by the failure to find extrachromosomal DNA in *S. lavendulae* and, in *Streptomyces kasugaensis,* where plasmids were demonstrated, no correlation was seen between the Arg$^-$ phenotype and loss of β-lactamase activity. However, Nakano et al.[247] noted an occasional spontaneous loss of plasmid from the wild-type *S. kasugaensis* and found that the resulting plasmidless strains generated Arg$^-$ derivatives at a much higher frequency than the parent. All Arg$^+$ revertants obtained from these had regained the plasmid, suggesting that, in this species, reversible integration of plasmid DNA into the *arg* locus does occur.

In another strain of *S. kasugaensis* producing pyrrothine antibiotics, marked variation in the complement of three plasmids was found in strains that arose spontaneously or by regeneration of protoplasts. Accompanying these variations were altered patterns of antibiotic production, sporulation, and citrulline auxotrophy. Chromosomal DNA from the wild type and some variants hybridized with plasmid DNA and it was tentatively suggested that the chromosome might contain a pleiotropic element into which plasmid DNA could insert, thereby activating a range of secondary insertions which caused the observed loss of functions.[265]

In *S. glaucescens*, the wild-type strain shows a moderate degree of sensitivity to streptomycin, but frequently produces variants with either increased or decreased sensitivity to the antibiotic. This variation appeared to be controlled by a gene *(strS)* that determines the natural level of resistance and that maps on the chromosome.[162] In the same oganisms, formation of melanin is also an unstable characteristic, giving rise at high frequency to nonproducing, tyrosinase-negative mutants. When pairs of such mutants were crossed, some of the combinations gave rise to melanin-producing progeny whereas other combinations did not. Analysis of the data revealed three groups of mutants that were presumed to have lesions in three separate genes, designated *melA, melB,* and *melC*. When representatives from each of these classes were crossed with auxotrophically marked *mel*$^+$ strains, *melA* and *melB* could be located on the chromosome. However, no fixed map location could be assigned to the *melC* gene. Restriction-enzyme analysis of the total DNA from *S. glaucescens* strains showed that a 9-Mdalton fragment was consistently absent from *melC* genomes and it was suggested that the loss might have resulted from a transposition event.[125] In one melanin-negative strain, a 7.3-kilobase (kb) DNA sequence that had been amplified to 500 copies was detected; this sequence has been isolated and cloned in *E. coli*.[280]

Schrempf[314] has carried out an extensive investigation into the causes of genetic instability in actinomycetes using the powerful biochemical techniques now available. In the initial studies with a *S. reticuli* strain that carried a plasmid of 48 to 49 Mdaltons, Amy$^-$, Mel$^-$, and Ant$^-$ (failure to produce the macrolide antibiotic, leukomycin) variants were found to arise spontaneously, and at higher frequency when treated with intercalating dyes.[312,315,317] Those that were Mel$^-$ but otherwise unchanged contained a plasmid approximately 1 Mdalton smaller than the wild type, and restriction endonuclease analysis indicated that the deletion had occurred in a 3-Mdalton *Sal*I fragment. Strains with more extensive phenotypic changes showed greater variation in plasmid DNA content. For example, strains that were Amy$^-$ Mel$^-$ and produced only small amounts of leukomycin contained a heterogeneous plasmid mixture of 9, 13, 44, 47, and 50 Mdaltons, while those that were Mel$^-$, Amy$^-$, and Ant$^-$ (Type V) completely lacked plasmid DNA. Although these results suggested that a *Mel* gene was present in the 3-Mdalton plasmid DNA fragment and that leukomycin production was also plasmid coded, the facility with which plasmid DNA rearranged raised questions about the possibility of concurrent chromosomal changes.

Restriction enzyme digests of total genomic DNA showed a series of 15 to 20 amplified fragments in all *Mel*$^-$ variants, the number and size distribution varying in individual colony isolates. Hybridization experiments established that these fragments were present without amplification in the wild type and were not associated with plasmid or ribosomal RNA genes. By nick-translating an amplified *Sal*I fragment from one of the Type V colonies and using it as a probe, several fragments in the reiterated DNA of each *Mel*$^-$ isolate were found to share homology. The hybridization pattern varied between isolates, and the shared homologies could also be demonstrated at low copy number in the wild-type strain. Moreover, amplified *Sal*I fragments of *S. reticuli* hybridized with amplified sequences in *Mel*$^-$ variants of *S. scabies* and other *Streptomyces* species. Schrempf[313] concluded that deletions, rearrangements, and amplifications could account for much of the variation in streptomycetes.

Studies on the actual location of the *Mel* gene were pursued[314] with a cloned 1.5-kb *Bcl*I DNA fragment from *S. antibioticus* that codes for tyrosinase, the sole enzyme necessary for melanin production.[175] Hybridization analysis established that *Mel* genes in *S. lusitanus, S. reticuli, S. scabies, S. azureus, S. antibioticus,* and several other Mel$^+$ strains contain extensive homologous regions.[314] Only *S. reticuli* and *S. lusitanus* contained detectable plasmids, and in these species the tyrosinase gene fragment hybridized with chromosomal

DNA and not with the plasmid. Furthermore, DNA from all *Mel*$^-$ derivatives of *S. scabies, S. antibioticus,* and *S. reticuli* failed to hybridize, regardless of their detectable plasmid status, indicating that the *Mel* gene had been deleted from the chromosome, or possibly from a large undetected plasmid. The results establish that rearrangements of DNA, whether spontaneous or promoted by plasmid-curing treatments, are responsible for loss of melanin production and that changes in any detectable plasmids present are incidental consequences of genetic instability.

It is apparent from numerous studies that one common result of treatments intended to cause plasmid elimination from actinomycetes is enhanced genetic instability. Although the mechanism by which this occurs is not yet known, distinct alterations in the electrophoretic pattern of fragments generated by digesting genome DNA with restriction endonucleases have been observed after treating cultures with intercalating dyes.[2,125,126,313] Where the formation and regeneration of protoplasts has been used to eliminate plasmids, disturbance of the membrane during protoplast formation may affect the stability of the genome. Although erroneous conclusions due to selection of spontaneous variants cannot be ruled out, several reports indicate that the protoplasting procedure can have appreciable genetic consequences. In one example, a strain of *S. fradiae* obtained by regeneration of protoplasts not only lost the ability to produce tylosin, but also acquired a tandemly amplified 10-kb DNA sequence.[23] A 300-fold tandemly repeated 6-kb sequence accounting for 15 to 18% of the total DNA has also been found in prototrophic recombinants regenerated after fusion of protoplasts from *S. jumonjinensis* and *S. lipmanii.*[299] The pattern of fragments containing the reiterated sequence varied in different recombinant strains and was not stable during storage of cultures. Hybridization studies indicated that the sequence originated in the *S. jumonjinensis* parent where it appeared to be involved in resistance to the antibiotic tunicamycin. Although events associated with protoplast fusion were presumed to have derepressed a system for repeated replication of a limited region of the genome, fusion per se may not have been responsible if amplification occurred while *S. jumonjinensis* protoplasts were being formed and regenerated, as observed in *S. fradiae.*

Apparent reiteration of a DNA sequence in a recombinational ''hotspot'' was encountered in plasmid pUC1061 after it was introduced by transformation into *S. espinosus.*[255] This plasmid is derived from the native *S. espinosus* plasmid, pUC6, and carries a 2.4-kb region of DNA that caused rapid plasmid loss from transformed strains of *E. coli* when it was present in shuttle vectors constructed from pUC6 and the *E. coli* plasmids pBR322 or pBR325. Although the ''hotspot'' was present within the 2.4-kb region causing shuttle-vector loss, it is not yet known whether the insertion phenomenon and vector instability are related in mechanism.

B. Mutagenesis

With mycelial organisms, mutagenic treatments are normally carried out with asexual conidiospores since these are uninucleate or uninucleoidal. If mycelium is used, a sporulation step must be interposed before selection.[250] Asporogenous organisms such as nocardia can be handled as highly fragmented vegetative mycelium.[320] Ultraviolet irradiation is a convenient mutagenic agent.[56,80,141] With *S. griseoflavus,* yields of auxotrophic mutants improved when the survivors were incubated on complete medium until sporulation occurred, then subjected to selection by allowing the spores to germinate in minimal medium and filtering to remove the prototrophic fraction.[44,307] Irradiating spores of *S. coelicolor* A3(2) and *S. scabies* with near ultraviolet (366 nm) light during exposure to 8-methoxypsoralen gave higher rates of mutation than were obtained with light of 254 nm.[355]

Of chemical mutagens that have been used, ethyl methanesulfonate is generally effective. *N*-Methyl-*N'*-nitro-*N*-nitrosoguanidine is a powerful mutagen when used under appropriate conditions.[68] It is not ordinarily suitable for producing mutants to be used in genetic studies because it gives a high proportion of strains with multiple mutations.[115] However, this property can be made use of for localized or directed mutagenesis[106,289] and, in conjunction with other fine-structure mapping procedures, for detecting adjacent genes by comutation.[290]

After repeated mutagenic treatment, yields of the desired mutants often decrease to unprofitable levels. The difficulty can usually be overcome by using a variety of mutagenic agents. With fungi, formation of a diploid can also lead to increased mutant recoveries from selected haploids when treatment is resumed.[213]

C. Plasmid Curing

To test the possibility that antibiotic biosynthesis is plasmid coded, producers have been treated with agents known to eliminate plasmid from *E. coli.*[128] These treatments have usually increased the proportion of non-producing variants in a population but, despite evidence from some studies that plasmids can be eliminated from *Streptomyces* species by curing agents,[124,154,317] the appearance of nonproducers is not necessarily associated with plasmid loss. Reports that other variable characteristics, such as the formation of aerial mycelium, sporulation, antibiotic resistance, A-factor synthesis, and pigment formation, are lost at higher than normal frequency on treatment with plasmid-curing agents, should also not be taken as firm evidence that the genes responsible reside on plasmids. The instability of such characters and of antibiotic production may often stem from genomic variations.[315] Various examples in which changes in antibiotic production have been recorded after exposure to heat, detergents, ultraviolet light, intercalating dyes, or novobiocin are discussed later (see Section VII).

D. Protoplasts

Techniques for forming protoplasts and regenerating vegetative structures have been developed for both unicellular and mycelial antibiotic-producing organisms.[18,152,272,284,309,339] The methodology varies with the biochemical composition of the cell walls. Protoplasts form readily from Gram-positive bacteria when the cells are incubated with lysozyme. The preparation of protoplasts from streptomycetes, first described by Sagara et al.[306] using *S. griseoflavus,* and later extended by Okanishi et al.,[272] is facilitated by growing cultures in a medium containing glycine at an optimum concentration, which predisposes the mycelium to disruption of its wall structure upon lysozyme treatment. As an alternative, germinating spores may be used.[302] The mycelial growth habit of streptomycetes usually regenerates well on a rich, high-osmotic medium[22,272,335] and can be promoted in strains that fail to give high yields of regenerants by adding a plasma expander.[164]

Formation and regeneration of protoplasts may be accompanied by loss of plasmids as a result of membrane effects.[255] Release of membrane vesicles has been seen to occur during protoplast formation[305,306] and the consequent potential removal of components required for replication and the partitioning of replicons during cell extension[169] could account for the elimination of extrachromosomal DNA. In *Streptomyces* species, plasmids are lost with a frequency of up to 20% in colonies obtained by plating protoplasts on regeneration medium.[136] The high rate of elimination has proven useful in studies with strains containing cryptic plasmids since progeny can be screened by direct examination of DNA extracts for plasmid loss.[139]

Protoplasting and regeneration of streptomycetes increases the range of variation in phenotypic characters. Regenerants of *S. ambofaciens, S. cirratus,* and *S. fradiae* included strains producing substantially increased titers of macrolide antibiotics, as well as variants altered in pigmentation and sporulation.[164] Similar observations have been reported for aureothricin production in regenerated protoplasts of *S. kasugaensis.*[265] Examination of antibiotic nonproducers present among regenerated protoplasts of *S. erythreus* suggested that chromosomal alterations had occurred since different specific steps in the pathway of erythromycin biosynthesis were blocked in each strain.[386] Changes in drug resistance have also been encountered.[388]

E. Screening for Increased Antibiotic Production

Detection of changes in antibiotic productivity is time-consuming since no generally applicable enrichment methods have been devised. Large numbers of colonies must be examined and procedures such as the agar-piece[73,163] or overlay[80] assays are commonly used in the initial qualitative screen. Promising strains are then evaluated quantitatively by growing cultures under conditions optimizing antibiotic production. Despite the effort required, this screening procedure can be used effectively if sufficient genetic variation is present in a culture. Mutagenic procedures that generate the necessary level of variation have provided enough progeny with enhanced productivity to meet the needs of industrial strain improvement programs.

The yield of high producers from the tedious antibiotic screen can be increased by conducting a preliminary selection for characteristics that correlate with high productivity. One such phenotype is superior resistance to the indigenous antibiotic.[82,374] As noted earlier (see Section IV.H.), a high antibiotic titer may limit further synthesis. This autoinhibitory effect can eventually frustrate attempts to increase productivity by repetitious mutation and selection. When this occurred in the aurodox producer, *S. goldiniensis,* selection of aurodox-resistant strains made possible the renewed isolation of superior producers.[362] If a group of antibiotics has the same resistance mechanism, as in the macrolide-lincosamide-streptogramin B series, any member can be used as the selective agent. Resistance in this group is due to methylation of 23S ribosomal RNA, and the methylase in producers is usually induced by the indigenous antibiotic. Since nonindigenous members of the series sometimes induce with lower efficiency, selecting for resistance to these weak inducers can give strains with a constitutive methylase and improved antibiotic production.[377]

Mutants with nutritional deficiencies tend to produce lower titers of antibiotic than their parent. Although the loss of productivity can sometimes be overcome by adding an excess of the required growth factor,[11] it is more often a pleiotropic effect of the mutation causing auxotrophy, at least in *P. chrysogenum,*[215] and can cause difficulties in genetic studies of antibiotic overproduction. However, interactions between auxotrophy and production may be used to advantage in targeted screening of mutants for increased titers. Reverted auxotrophs show more variation in productivity than the wild-type parent, and the proportion of superior producers is usually favorable where the auxotrophic requirement involves a precursor of the antibiotic. Knowledge of the biosynthetic pathway is useful here and in making possible the selection of mutants resistant to analogs of a primary precursor of the antibiotic, which usually include a high proportion of overproducers.[71]

Selection of ''revertant'' producing strains after further mutagenic treatment of nonproducing mutants often yields populations highly enriched with producers superior to the original grandparent, thereby increasing the efficiency of screening procedures. The overproducers are presumed to be regulatory mutants since success varies with the particular nonproducing derivative chosen for reversion.[80]

VII. LOCATION OF GENES FOR ANTIBIOTIC PRODUCTION

A. Plasmid Genes

The possibility that antibiotic production might be under the control of one or more plasmid genes offers a plausible explanation for many of its characteristics.[222] Noteworthy among these is a tendency for producing strains to ''run down''. Low-productivity variants, often with no obvious differences in growth characteristics, appear in the population too frequently to be due to random mutations. Antibiotic production is also an inconstant property of species in their natural habitat, suggesting that it is not essential for survival and that, under circumstances where its loss increases competitiveness by permitting faster growth, the biosynthetic process might be discarded by eliminating a plasmid. That plasmids might be involved in secondary metabolic activities was first suggested from studies of melanin biosynthesis by *S. scabies.*[111-113] Later, Okanishi and co-workers observed an accelerated loss of aerial mycelium and antibiotic productivity when cultures of *S. kasugaensis* were treated with plasmid-curing agents.[267,271] Numerous reports of similar behavior have since appeared in the literature.

The criteria that should be used to assess whether a trait such as antibiotic production is plasmid controlled have been discussed by Hopwood[134] and Okanishi.[264] The principal considerations are (1) genetic evidence for infectious transfer of the phenotype at a frequency several orders higher in magnitude than chromosomal recombination, (2) absence of linkage between genes for the trait and known chromosomally coded traits, (3) instability of the phenotype and a marked decrease in stability through curing, and (4) correlation with the presence (and loss on curing) of extrachromosomal DNA. Despite considerable efforts to confirm preliminary indications of plasmid involvement in antibiotic production, only the biosynthesis of methylenomycin can be considered to satisfy these criteria.

1. Methylenomycin

Infectious transfer of the ability of *S. coelicolor* A3(2) to produce an inhibitory substance has been correlated with transfer of the postulated fertility plasmid SCP1.[181] The inhibitory substance has been characterized as methylenomycin.[384] A variety of genetic tests, including transfer of inhibitor production to *S. parvulus* and *S. lividans* 66 and the preparation in an SCP1′-*cysB*$^{+}$ strain of 16 nonproducing *(mmy)* mutants that all showed linkage to the plasmid, support the conclusion that genes for the biosynthesis of methylenomycin and for resistance to the antibiotic are located on SCP1.[180] Hornemann and Hopwood[156] have reviewed this evidence and the attempts made to detect plasmid DNA carrying the genetic traits associated with SCP1.

Methylenomycin was first isolated from *S. violaceoruber* SANK95570 in which a 110-Mdalton plasmid, pSVl, was subsequently detected.[270] The presence of pSVl in this organism correlates with methylenomycin production.[1] When DNA specifying methylenomycin resistance, obtained by shotgun cloning in *E. coli*[34] of the total DNA from an SCP1^{+} strain of *S. coelicolor* A3(2), was used as a radioactive probe, it hybridized specifically with pSV1. Plasmid pSV1 could be transferred conjugatively into SCP1^{-} strains of *S. coelicolor* A3(2) or *S. lividans,* after which the recipients exhibited methylenomycin production and resistance; similar results were obtained by transforming these strains with pSV1 DNA.[137] Thus, although the evidence for plasmid-directed biosynthesis of methylenomycin in wild-type *S. coelicolor* A3(2) remains incomplete until SCP1 is isolated, no such reservations apply to locating the relevant *S. violaceoruber* SANK95570 genes on pSV1.

B. Reported Plasmid Involvement

There are numerous reports linking antibiotic production to plasmids, or behavior considered indicative of plasmids, in actinomycetes and other bacteria. Much of the evidence rests on a single criterion, such as an increased incidence of nonproducing strains after treatment with plasmid-curing agents, or indications of infectious transfer of antibiotic producing ability, and its interpretation is now open to question. The nature of this evidence for individual antibiotics is discussed in the following sections.

1. Macrolide Antibiotics

Plasmid involvement in tylosin biosynthesis was suspected from the recovery of 20 to 60% nonproducers among recombinants formed during protoplast fusion of auxotrophically marked producing strains of *S. fradiae*. The proportion of nonproducers was similar for all markers transferred[19,20] and was subsequently shown to occur during formation and regeneration of protoplasts rather than during recombination. A nonproducer selected from the regenerants and genetically marked with spectinomycin resistance recovered the ability to make tylosin when mixed with producer strains under conditions where no chromosomal recombination was detected.[23] Although the results implicate a transmissible plasmid as the carrier of antibiotic production genes, chromosomal DNA seems also to be required since the *tylA* gene, which controls the formation of sugars in the tylosin biosynthetic pathway, shows linkage to auxotrophic markers.[19]

In *S. reticuli*, loss of leukomycin production was correlated with elimination of a 48- to 49-Mdalton plasmid by treatment with acriflavine or ethidium bromide,[312] but since chromosomal DNA was also extensively altered,[313] this offers no compelling reason to implicate the plasmid in antibiotic production. Admixture of prototrophic plasmidless strains with auxotrophic wild-type strains containing the plasmid resulted in 50 to 100% plasmid transmission into the auxotroph and, as in the tylosin producer, there was a concomitant reappearance of antibiotic synthesis. However, attempts to confirm the plasmid role by restoring the wild-type phenotype through transformation of plasmidless prototrophs with plasmid DNA were unsuccessful and the evidence for plasmid involvement in leukomycin production remains equivocal. Examination of other macrolide producers showed extrachromosomal DNA to be present in strains of *S. albogriseolus* and *S. griseoflavus*, and in some strains of *S. ambofaciens* and *S. erythreus*, but absent or probably absent from other strains of the latter and also from *S. violaceoniger* and *S. tendae*.[311] A survey of macrolide-producing actinomycetes conducted by Omura et al.[277] showed plasmid DNA in only 6 of 22 streptomycetes examined. Thus, even where plasmids are found in a species, their presence is strain specific and does not correlate with antibiotic production.

Cultures of *S. ambofaciens* ISP5053, which produces spiramycin, contain 1 to 10% of nonproducing natural variants; this relatively high frequency is not appreciably altered by treatment with acriflavine.[276] A 53-Mdalton plasmid, pSA1, present in the wild type, could still be detected in the nonproducing strains although detection was difficult because of an increased association with membrane material.[277] Elimination of pSA1 gave a strain that was unable to produce spiramycin, but partially recovered this activity after protoplast formation and regeneration and fully recovered it when pSA1 was reintroduced by conjugation with a plasmid-containing strain. From these results, Ikeda et al.[165,166] concluded that pSA1 enhanced spiramycin excretion but did not contain genes for the biosynthetic pathway. Support for the latter conclusion has been supplied by Schrempf,[311] who found no homology when DNA digests from the *S. ambofaciens* plasmid were hybridized with DNA from the plasmid of *S. reticuli*. Since biosynthesis of the aglycone is expected to be common to both producers, this suggests that its genes are not present on either plasmid.

S. erythreus NRRL2338 is unexpectedly stable to plasmid-curing treatments with ethidium bromide, acridine orange, novobiocin, nalidixic acid, and rifampicin. With ethidium bromide,[376] the 1 nonproducer present in 2000 colonies tested after exposure to the dye appeared to be blocked in a specific biosynthetic reaction. Formation and regeneration of protoplasts caused loss of erythromycin production with a frequency of only 0.13% and the nonproducers also each blocked at a specific step in the biosynthetic pathway. To explain these results in terms of plasmid elimination, the pathway genes would have to be distributed among a number of discrete DNA elements and, indeed, *S. erythreus* NRRL2338 was found to contain a complex array of nine plasmids ranging in size from 7 to 94 Mdaltons.[375] The results from genetic crosses between Ery^+ and Ery^- strains carrying chromosomal markers show a marked similarity to those obtained for tylosin[23] and leukomycin[311] production in that all of the progeny isolated were producers.[376] Although selection against antibiotic-sensitive nonproducing progeny cannot be excluded, this indication of infectious transfer supports a plasmid location for erythromycin production genes.

Of the other macrolide antibiotic producers, *S. hygroscopicus* loses the ability to produce turimycin at a high rate during treatment with ethidium bromide or acridine orange[172] and two *Streptomyces* species producing lankacidins contain closely related linear DNA plasmids, pSLA1 and pSLA2, which associate with protein at their 5′-ends.[127] The 11-Mdalton pSLA2 was estimated to have a copy number of 60 in *Streptomyces* sp. 7434-AN. Treatment of this strain with ethidium bromide gave nonproducing variants at a frequency of 4.4%. When one of these was examined, it was found to have lost its natural resistance to lankacidins and to have also lost the plasmid.[124]

2. Chloramphenicol

Early genetic evidence suggested that genes responsible for chloramphenicol production in *S. venezuelae* ISP5230 are distributed between the chromosome and a plasmid.[3,4] Crosses between strains bearing chromosomal markers and those with markers for structural genes in the pathway of antibiotic biosynthesis invariably showed linkage, whereas plasmid control was indicated by a high-frequency loss of production during treatment with intercalating dyes[242,267] and by the results of genetic crosses when dye-treated nonproducing strains were used.[267] In the latter crosses, the Cpp marker could not be mapped on the chromosome. However, no plasmids can be detected in *S. venezuelae* ISP5230, nor in several other chloramphenicol-producing strains, and two strains, *S. venezuelae* 13s and *S. phaeochromogenes* NRRLB3559, that contained plasmids of 17 and 7 Mdaltons, respectively, retained full chloramphenicol productivity after the extrachromosomal DNA had been eliminated by formation and regeneration of protoplasts.[2,77] If any genes specifying chloramphenicol production are extrachromosomal, the plasmid is not detectable by methods currently in use.

3. Aminocyclitol Antibiotics

Streptomycin production, and resistance to the antibiotic, was lost in 2 to 16% of *S. bikiniensis* ATCC11062 colonies after acriflavine or ethidium bromide treatment,[332] and in *S. griseus,* a nontransmissible plasmid has been implicated in the formation of streptomycin.[386] The latter plasmid also carried genes for natural resistance to lincomycin and could be transferred into a nonproducing plasmidless strain by exposing calcium-treated cells to plasmid DNA.[396] Transformation was monitored by testing the sensitivity of the recipient to lincomycin; 66% of the resistant progeny produced streptomycin.

Plasmid-curing treatment of *S. fradiae* ATCC10745 yielded two types of strains that had lost the ability to produce neomycin.[387] One was sensitive to exogenous neomycin, probably due to a much reduced level of 3′-*O*-phosphotransferase [APH(3′)], and showed no response to supplements of 2-deoxystreptamine, an intermediate in neomycin biosynthesis. In the other type, which retained its natural resistance to the antibiotic, neomycin production was restored by 2-deoxystreptamine. Although these results suggested that two plasmids, one responsible for neomycin resistance and the other for the biosynthesis of 2-deoxystreptamine, were involved, only a single plasmid of 52 Mdaltons could be found in wild-type mycelium.[187] Moreover, in a modified 56-Mdalton form, the plasmid was still present in a neomycin-sensitive nonproducer from the ethidium bromide treatment. Hybridization studies using cloned neomycin resistance genes of *S. fradiae*[353] as probes failed to locate the resistance determinants on any of the plasmids and it was proposed that alterations in chromosomal DNA were responsible for the changes.

Loss of kanamycin production in *S. kanamyceticus,* which occurs spontaneously at a relatively high (0.2 to 1%) frequency and is associated with a change in colonial morphology,[57] is increased many-fold by treatment with intercalating dyes or incubation at increased temperature. The nonproducers are more sensitive to kanamycin or amikacin, but retain much of the genetic information for kanamycin biosynthesis since they are able to produce the antibiotic when supplemented with 2-deoxystreptamine, streptamine, or glucosamine. Variants with this phenotype have been observed in several strains of *S. kanamyceticus*[159] and their formation has been taken to indicate plasmid involvement in kanamycin production.[82,159]

In *S. ribosidificus,* preliminary evidence suggesting plasmid involvement in ribostamycin production was obtained from dye and heat treatments similar to those described for kanamycin. Further investigation showed that a plasmid was present in the wild-type producer but it does not appear to control ribostamycin production since it was still present, apparently unaltered, in the nonproducing variants thought to have been cured by acriflavine treatment.[253]

4. Pyrrothine Antibiotics

From the increased frequency with which thiolutin, kasugamycin, and aureothricin production were independently lost after treating *S. kasugaensis* M338 with acriflavine, plasmid-borne genes were postulated to be involved.[271] A second *S. kasugaensis* strain, MB273, which sporulated better and also produced the antibiotics, was subsequently shown to harbor three plasmids: pSK1 (6.8 Mdaltons), pSK2 (6.8 Mdaltons), and pSK3 (14.5 Mdaltons).[263] Restriction enzyme mapping established that pSK3 was a composite of pSK1 and pSK2.[265,356] One spontaneously formed nonproducer in which all three plasmids were present but with the relative abundance of pSK3 greatly increased, was able to form aureothricin when supplemented with cystine, a precursor of the antibiotic, but also responded to palmitic or oleic acids, suggesting a possible membrane change. Among several variants present in regenerated protoplasts of this strain were some from which pSK1 was absent and others that had lost all three plasmids, but both types produced aureothricin on agar medium. Since hybridization experiments established that neither pSK1 nor pSK2 had integrated into the chromosome, the genes specifying antibiotic biosynthesis could not have been located on these plasmids and may be chromosomal;[98] the pleiotropic changes observed are presumably due to genetic instability in *S. kasugaensis,* which is increased by plasmid-curing treatments.

With another pyrrothine producer, *S. clavuligerus,* Kirby[179] observed loss of holomycin production in agar cultures when spores were irradiated with ultraviolet light. The incidence of Hol^- strains reached 2.9%; in contrast, acridine orange or nitrosoguanidine were relatively ineffective. When chromosomally marked Hol^+ and Hol^- strains were grown in mixed culture, the holomycin production trait was transferred nonselectively at frequencies up to 23%. In prototrophic recombinants, this frequency was much higher, ranging from 45 to 93%. Attempts to place the Hol^- character on a map of auxotrophic markers[178] failed to give a fixed map position, whereas *hol*$^-$ marker obtained by nitrosoguanidine treatment showed an unambiguous location between the closely linked *ade*-1 and *phe*-1 markers. Other evidence also supported the idea that the Hol^- strains induced by ultraviolet irradiation were altered in regulatory rather than structural genes and it was postulated that *S. clavuligerus* contains an unstable self-transmissible genetic element, SUE 1, that controls production of holomycin. Attempts to detect extrachromosomal DNA in the organism gave negative results.

5. Actinomycin

In the actinomycin-producing organisms, *S. antibioticus* and *S. parvulus,* exposure to acriflavine or novobiocin increased the frequency with which nonproducing strains appeared without causing corresponding increases in the number of morphological or auxotrophic variants.[256] This indication of plasmid involvement was supported by the recovery of predominantly (84 to 95%) actinomycin-producing prototrophs after protoplast fusion of auxotrophically marked parental and dye-treated nonproducing strains, whereas only nonproducing recombinants were recovered when nonproducing strains were fused.[257] However, this does not take into account the likelihood that nonproducing progeny would be selected against because of their sensitivity to actinomycin excreted by the producers.

6. Gramicidin S

Plasmids in gramicidin S-producing strains of *Bacillus brevis* have also been found in the nonproducers that are formed at high frequency during treatment with intercalating dyes.[75,226] They are not directly implicated in gramicidin S production. In *B. brevis* ATCC9999, the plasmid varies in size (45 to 47 Mdaltons) when associated with membrane or chromosomal material, but restriction enzyme analysis indicated that the DNA in the different forms is closely related.[227] Dobritsa et al.[75] detected repeat sequences in plasmid pAD1 from *B. brevis* strain G. B. and surmised that the variability in antibiotic production might be due to genetic instability of the kind promoted by transposable elements.

7. Nisin

Streptococcus lactis strains that produce the oligopeptide antibiotic, nisin, lose this property at frequencies up to 1% when exposed to plasmid-curing treatments.[189] Extrachromosomal DNA present in the producing parents was absent from some of the nonproducing derivatives but could still be detected in others so, unless the curing treatment had caused alterations in the residual plasmids, it is doubtful that nisin biosynthesis is directed by the plasmid in question.[95]

8. Other Antibiotics

A high frequency of nonproducers in cultures treated with plasmid-curing agents has also been observed with *Streptomyces* sp. M-506, which makes antimycin,[84] and with the puromycin producer, *S. alboniger*.[308] In addition, plasmid-curing treatments are reported to accelerate the loss of oxytetracycline resistance in the producer, *S. rimosus*,[41] of tetracycline resistance, pigmentation, and sporulation in another streptomycete,[46] and of A-factor production in *S. griseus*.[117] In each case, plasmid involvement has been suggested, but only for A-factor production has strong supporting evidence been presented. Hara et al.[119] were unable to assign a fixed map position to markers for A-factor biosynthesis by analyzing recombinants recovered after protoplast fusion of *S. griseus* strains. In contrast, *afs* genes for A-factor production in *S. coelicolor* A3(2), where the phenotype is stable to plasmid-curing agents, mapped unambiguously on the chromosome. In addition, the cloned *afsA* gene from *S. coelicolor* A3(2) failed to hybridize with DNA from most *S. griseus* strains that were deficient as a result of curing treatments.[154] Infectious transfer from producing to nonproducing strains of the capacity to make the enzyme-inhibiting peptide, leupeptin, during conjugation between strains of *S. roseus* MP839-A1 likewise suggests that a plasmid may carry genes for the production of this secondary metabolite,[361] but a report[360] that *S. coelicolor* A3(2) produces an antibiotic, for which genetic determinants are carried by the plasmid SCP2, has been questioned.[137]

C. Chromosomal Genes (Actinomycetes)

In some actinomycetes, genes for antibiotic production have been convincingly shown to lie on the chromosome. The evidence comes primarily from analysis of linkage between markers for antibiotic production and auxotrophic or resistance markers. Although no fine-structure mapping has yet been carried out, there are good indications that genes for some antibiotic biosynthetic pathways are clustered. Sermonti and Lanfaloni[325] analyzed linkage data for a number of *Streptomyces* and for *N. mediterranea* by aligning similar loci in a circular map. The genes for antibiotic pathways not only showed a noticeable clustering, but were also most often found in a similar region of the chromosome. It is noteworthy that antibiotic production shown to be chromosomally determined is no less unstable than that in which plasmid-borne genes have been suspected to play a role.

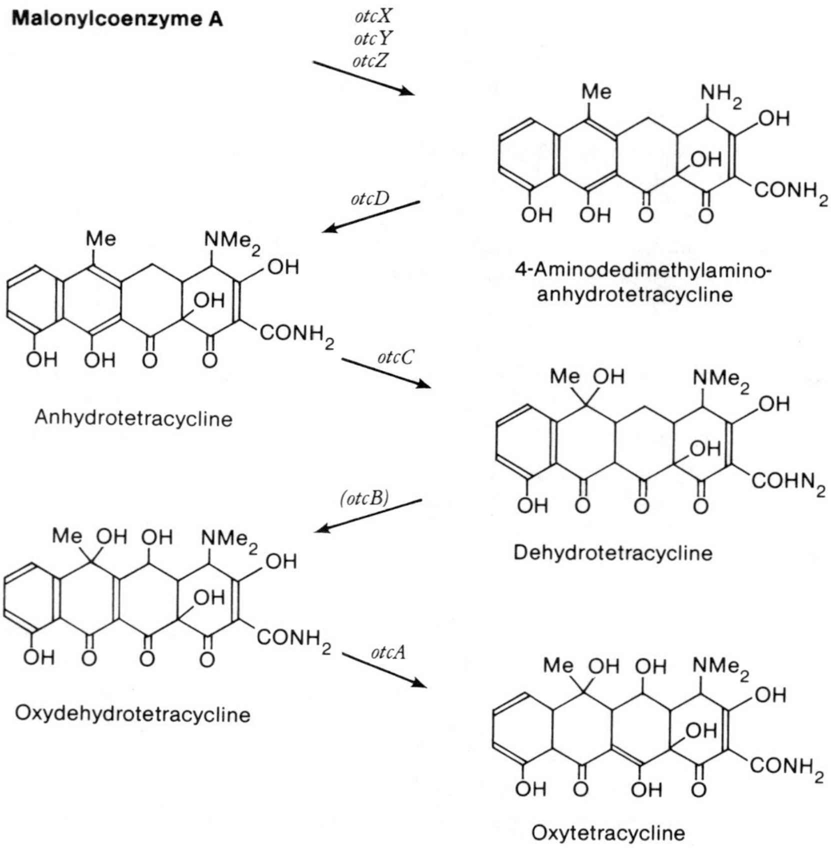

FIGURE 1. Partial sequence of reactions in the biosynthesis of oxytetracycline by *S. rimosus*

1. Tetracyclines

The pathway for tetracycline biosynthesis is well defined biochemically (Figure 1). Much of the information that allowed the correct biosynthetic sequence to be distinguished was obtained through the use of blocked mutants; in *S. aureofaciens,* which produces mainly tetracycline and chlortetracycline, 11 steps could be recognized and placed in sequence by cosynthesis tests.[238] Mutations that caused loss of chlortetracycline production mapped at two locations in the region between a pair of different loci for streptomycin resistance.[40] Although recombination has been demonstrated,[9,38] definitive information on the genetic map of this species has not yet been published and the precise location of the genes is uncertain. The situation is considerably better for oxytetracycline biosynthesis in *S. rimosus,* where detailed linkage maps of genetic markers have been developed.[6,91] Early studies by Alikhanian et al.[12] yielded two classes of nonproducing mutants, characterized by a ''white'' or ''black'' colony appearance, that produced oxytetracycline when grown in black-white combinations. A subsequent study of more than 70 nonproducing mutants also gave two cosynthesis groups,[243] and analysis of recombinants between genetically marked strains indicated that one group *(otcA)* was linked to *strA* while the other *(otcB)* was associated with a different group of markers.

Later work on *otc* mutants defined four cosynthesis groups which mapped in the region between *his* and *phe*.[5,286] Recently, with a different strain of *S. rimosus,* seven distinct types of nonproducing mutants were identified.[40] These were grouped in two classes; those in class 1 lacked cosynthetic factor 1, a deazaflavin cofactor required for the final step in the

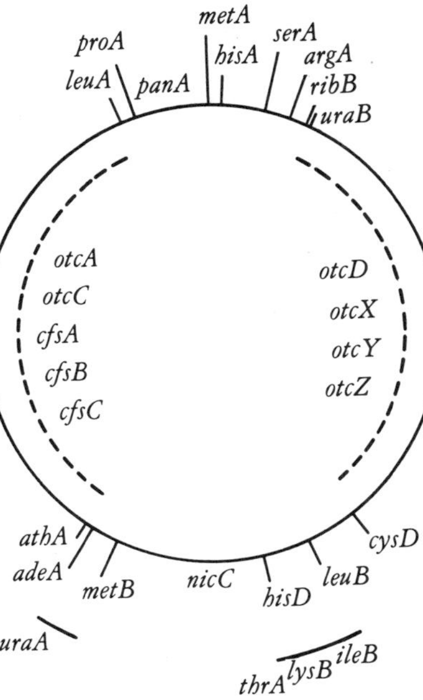

FIGURE 2. Location of genes for oxytetracycline synthesis on the linkage map of *S. rimosus*. Marker positions have been arbitrarily related to those of *S. coelicolor* A3(2).

biosynthesis of tetracyclines, and were considered to have mutated *csf* genes. Lesions in class 2 strains were assigned by precursor supplementation and cosynthesis tests to various steps in the oxytetracycline pathway itself (Figure 1). No strain with a defective *otcB* gene was found and the sequence of reactions for the *otcX, otcY,* and *otcZ* genes could not be determined. Genes for the final sequence of reactions converting anhydrotetracycline to oxytetracycline (*otcA* and *otcC*) were located close together with the *csf* genes in the region of the genetic map between *proA* and *adeA* whereas those specifying the biosynthetic reactions preceding 4-amino-dedimethylaminoanhydrotetracycline in the biosynthetic pathway were placed on the opposite side of the chromosome, between *ribB* and *CysD* (Figure 2).[296]

2. Actinorhodin

S. coelicolor A3(2) produces several different antibiotics. In SCP1$^-$ strains, which cannot produce methylenomycin, some of the antibacterial activity is associated with actinorhodin, an indicator pigment of polyketide origin found in cultures identified as *S. coelicolor, S. lividans,* and *S. violaceoruber.*[42,48,153] Wright and Hopwood[385] isolated five Act$^-$ mutants in SCP1$^-$ strains carrying auxotrophic and streptomycin-resistance markers. When these were crossed with auxotrophically marked producing strains, analysis of recombinants indicated that the *act* genes were linked to the chromosomal markers and mapped between *cysD* and *strA*. Differences in the mutant phenotypes indicated that the lesions were in different genes and the mapping data suggested a gene cluster. In a more extensive study, 76 mutants blocked in actinorhodin production were isolated and classified into phenotype groups I to VII.[303,304] Cosynthesis tests indicated that the reactions coded by their genes were arranged in the sequence:

$$\xrightarrow{\text{I, III}} \xrightarrow{\text{VII}} \xrightarrow{\text{IV}} \xrightarrow{\text{VI}} \xrightarrow{\text{V}}$$

Class II strains failed to complement those of other classes in any combination and may

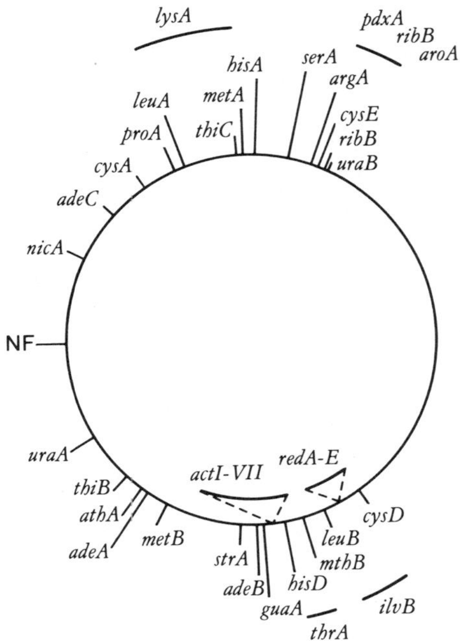

FIGURE 3. Location of genes for actinorhodin and undecylprodigiosin biosynthesis on a partial linkage map of *S. coelicolor* A3(2).

have been regulatory mutants. The genes for all of these mutations mapped on the chromosome in the narrow region between *guaA* and *hisD* (Figure 3). Clustering of genes specifying the complete actinorhodin pathway was subsequently confirmed by cloning a 59-kb segment of the chromosomal DNA (see Section VII.F).[223]

3. Prodigiosin

A second red-colored antibiotic in *S. coelicolor* A3(2) has been tentatively identified as undecylprodigiosin.[304] In a type-I *act* strain, where no actinorhodin was made and the second pigment could be unambiguously detected, mutants blocked in undecylprodigiosin formation (designated *red*) could be grouped by cosynthesis tests into five classes (A to E). The *redB*, *redC*, and *redE* strains each secreted intermediates that *redA* converted to undecylprodigiosin, and a mixture of *redB* and *redC* likewise produced the pigment, but the low diffusibility (or instability) of intermediates prevented a more detailed ordering of the sequence. Analysis of recombinants in crosses with chromosomally marked strains established that the genes for antibiotic production were located on the chromosome. All of the *red* markers mapped close together in the region between *leuB* and *lysD*, well separated from the *act* cluster (Figure 3). Evidence that the genes are closely linked has subsequently been supported by cloning and hybridization studies (see Section VII.F).

In *S. marcescens*, the absence of detectable plasmids in some strains, and the nonidentity of restriction enzyme fragments from the Hy plasmid present in other producers with fragments similarly obtained from cloned DNA containing prodigiosin pathway genes, suggests that biosynthesis of the antibiotic is chromosomally directed in this organism also (see Section VIII.F).[66]

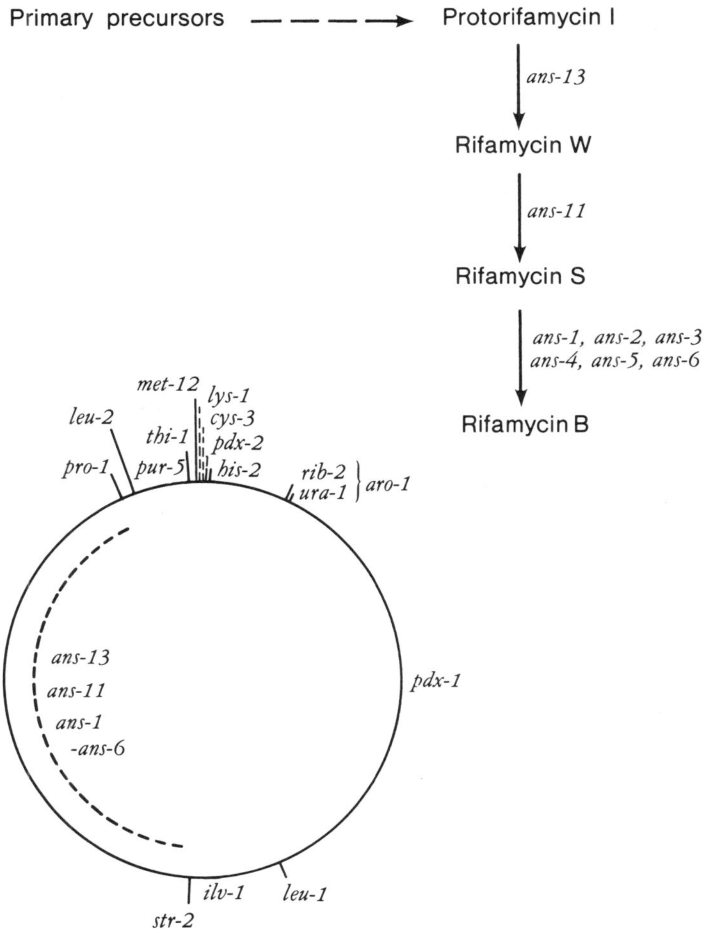

FIGURE 4. Partial sequence of reactions and genetic map of genes in the pathway of rifamycin biosynthesis in *Nocardia mediterranea*. Marker positions have been arbitrarily related to those of possible corresponding markers in *S. coelicolor* A3(2).

4. *Rifamycin*

Mutations causing loss of rifamycin B production in *N. mediterranea* showed clustering on the chromosome. A mutation *(ans-13)* in the steps between protorifamycin I and rifamycin W and another *(ans-11)* between rifamycin W and rifamycin S mapped together in the region between *pro-1* and *str-2*. A group of mutants blocked in the conversion of rifamycin S to rifamycin B and designated *ans-1* to *ans-6* also mapped in the same region (Figure 4).[103,319]

5. *Zorbomycin-Zorbonomycin-Antibiotics*

This group of peptide-type antibiotics is produced by *S. bikiniensis* var. *zorbonensis* for which a genetic map of 14 auxotrophic and resistance markers has been prepared by Coats and Roeser.[65] Nonproducing strains carrying antibiotic pathway markers were obtained by mutagenesis with *N*-methyl-*N'*-nitro-*N*-nitrosoguanidine; two different mutations (*zorA* and *zorB*) prevented all antibiotic synthesis whereas a third *(zorD)* eliminated only the formation of zorbonomycins B and C. The *zorA* and *zorB* mutations mapped together in the region close to *metB* and distant from *zorD* (Figure 5).

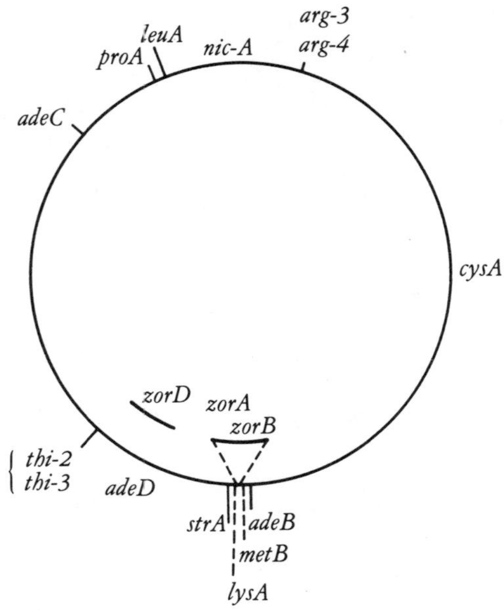

FIGURE 5. Relative location of genes involved in biosynthesis of the zorbomycin-zorbonomycin group of antibiotics on a linkage may of *S. bikiniensis* var. *zorbonensis*. Marker positions have been arbitrarily related to those of *S. coelicolor* A3(2).

6. Other Antibiotics

Three antibiotics included in the preceding section because of reported plasmid involvement have been shown to have structural biosynthetic genes linked to chromosomal markers. These are holomycin,[179] tylosin,[19] and chloramphenicol.[4]

D. Chromosomal Genes (Fungi)

1. Penicillin Biosynthesis

Extensive studies of penicillin biosynthesis have provided a relatively detailed understanding of the pathway (Figure 6). This biochemical background has assisted genetic analysis[248] and explained such observations as the loss of penicillin production in 25% of the lysine auxotrophs of *P. chrysogenum*.[39] The genes for penicillin production are chromosomal. Early evidence that they are located in the nucleus and not in the cytoplasm was provided by constructing heterokaryons of genetically marked producer and nonproducer strains.[216,322] Normansell et al.[254] isolated 78 nonproducing *(npe)* mutants and classified them by complementation tests into five groups (V, W, X, Y, and Z). Most fell in group Y, which biochemical tests indicated to be blocked at a stage preceding the Arnstein tripeptide, δ-(α-aminoadipyl)cysteinyl-D-valine.[207] Groups X and Z were also blocked in the early segment of the pathway whereas the lesion in group W was between the Arnstein tripeptide and isopenicillin N; in group V it was between isopenicillin N and penicillin G. Haploidization analysis of recombination in the parasexual cycle showed that the mutant groups W, Y, and Z were associated on one, group X on the second, and group V on the third of the three chromosomes believed to be present in *P. chrysogenum*.[81,219]

Mutants of *A. nidulans* blocked in penicillin biosynthesis have been classified by diploid complementation tests into four groups designated *npeA*, *npeB*, *npeC*, and *npeD*, with the largest number falling into the *npeA* group.[130] All *npeA* mutants regained the ability to produce penicillin when supplemented with the Arnstein tripeptide, suggesting that the block

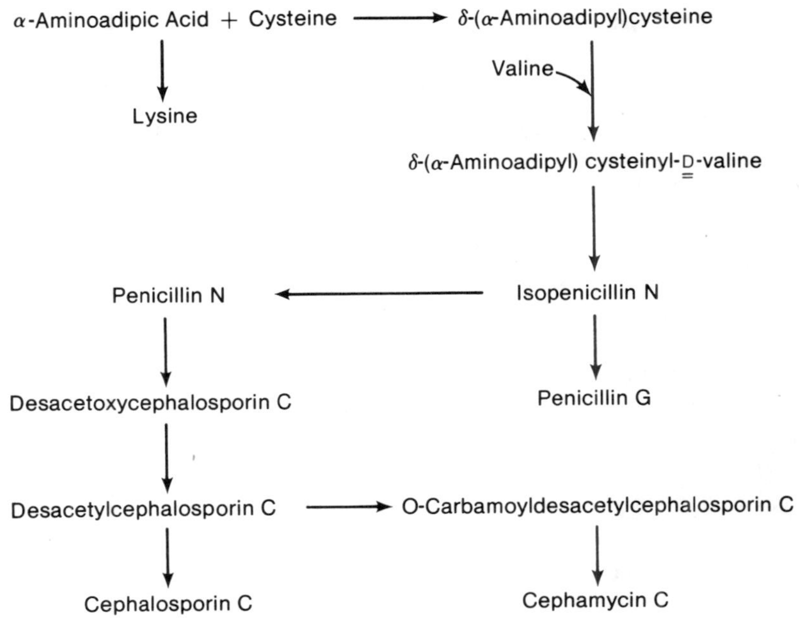

FIGURE 6. Biosynthetic relationships between lysine, penicillins, cephalosporins, and cephamycins.

lies in one or more steps leading to this intermediate, and that this group corresponds to the *npeY* group of *P. chrysogenum.*[212] Eight chromosomes are present in *A. nidulans* and all *npeA* mutants appear to be located on chromosome VI. The *npeB, npeC,* and *npeD* mutants were located on chromosomes III, IV, and II, respectively. More extensive genetic analyses have positioned the *npe* genes in relation to other markers on the respective chromosomes.[212]

2. *Penicillin Overproduction*

Mutations responsible for increasing penicillin production in *P. chrysogenum* are recessive in heterozygous diploids.[216] Thus, although diploids can be formed without difficulty via the parasexual cycle,[217] they do not offer an easy route to strains with superior productivity and few such heterozygotes have been described. On the other hand, where a trait is dominant in the phenotype, heterozygous diploids can be used to advantage. In one construction, a haploid giving high penicillin titers but sporulating poorly was combined with another of the same mutagenized lineage that produced less antibiotic but sporulated well. Selected diploids synthesized penicillin at the level of the less productive parent but, because sporulation is dominant, produced abundant spores.[55] Penicillin titers were subsequently increased by selection after treatment with nitrogen mustard.[212] Construction of homozygous diploids from haploids that differ only in neutral markers introduced to detect diploidy can also raise penicillin production above that of the parents.[83] This is attributed to mutual masking of highly mutable recessive genes that give rise to deleterious mutants; the accumulation of these during inoculum build-up and growth of fermentation cultures is presumed to limit haploid productivity.[212]

The advantage of improved diploid strains can be negated by a high rate of segregation, causing high-producing heterozygotes to lose the beneficial effects of dominant genes as inferior haploid phenotypes accumulate during a fermentation cycle. Genetic strategies have been devised to ensure that the heterozygous state is retained. Macdonald[210] proposed that nonallelic recessive mutations causing slow growth be introduced in each haploid parent, or alternatively, that auxotrophs showing growth retardation in complex media be used as the haploid parents. Mutagenesis and selection of nonallelic recessive lethals in the diploid have also been suggested.[14,16]

Haploid strains of *P. chrysogenum* with improved penicillin production have been synthesized by exploiting the recombination mechanisms of the parasexual cycle.[8,16] However, recombination was found only in diploids constructed from closely related strains.[218] Its absence when parental types differed appreciably in lineage did not result from intracytoplasmic selection favoring the parental genome and was attributed to genetic incompatibility due to an accumulation of mutagen-induced chromosomal changes.[211]

Penicillin overproduction in *P. chrysogenum* has been analyzed genetically by constructing diploids from strains of different productivity and examining haploid recombinants formed by parasexual mechanisms. Ball was able to assign four mutations affecting penicillin yield to haploidization group I; a fifth was tentatively assigned to group III.[15] Only crosses between the group III and group I strains yielded recombinants with higher production than either parent.[17]

Although the usefulness of *A. nidulans* for genetic analysis of penicillin overproduction is hampered by its low productivity, some valuable insights have been obtained. A heterokaryon test combining three high-producing mutants, $penA_1$, $penB_2$, and $penC_3$, with a low-producing strain showed that high productivity was invariably, as in *P. chrysogenum,* a nuclear trait. However, the $penB_2$ mutation was dominant and $penC_3$ was semidominant; only $penA_1$ was recessive.[74] By parasexual haploidization analysis, the three mutations were located on chromosomes VIII, III, and IV, respectively.[214] Parasexual crosses with pairwise combinations of the three strains yielded haploids combining the positive traits, but none showed superior production to the parents.[74,213] Similar results were obtained in sexual crosses or when high-producing haploid strains selected after several round of mutagenic treatment were used to produce a diploid.[336] Difficulties in constructing superior homozygous diploids from haploids carrying the same positive mutations are ascribed to the additive effects of accompanying deleterious mutations cancelling the higher yield expected from doubling the positive effects. Recognizing this, and the possibility that it might be avoided by obtaining disomics in which only the chromosome carrying the positive mutation is duplicated, Upshall et al.[363] have described a method for obtaining stable disomic strains of *A. nidulans.*

The $penA_1$, $penB_2$, and $penC_3$ mutations have been positioned in relation to other markers on the same chromosome by analyzing recombination frequencies in appropriate crosses.[212] Such analyses, involving the quantitative estimation of penicillin production, pose considerable technical problems, as well as difficulties in interpretation of data.[213]

3. Cephalosporin Biosynthesis

The biosynthesis of cephalosporins and their relationship to the penicillin family of β-lactams is shown in Figure 6.[70]

Mutants of *A. chrysogenum* blocked in cephalosporin C production could be divided into two groups, depending upon whether the lesion occurred before or after formation of the Arnstein tripeptide.[198] Some heterokaryons formed when strains in the early lesion group were mixed in pairwise combinations recovered the ability to produce cephalosporin, confirming that several genes specify this section of the pathway. Among the late-pathway mutants, three groups could be distinguished by their ability to accumulate only penicillin N, penicillin N and desacetoxycephalosporin C, or a mixture of these with desacetylcephalosporin C.[96,97,200,288]

4. Cephalosporin Overproduction

As a rule, cells in the vegetative mycelium of *A. chrysogenum* are uninucleate, making it difficult to obtain heterokaryons and to synthesize diploid strains by natural processes. In a study by Hamlyn and Ball,[116] two divergent high-producing lines (A and B) obtained by repeated round of mutagenesis and selection from a low-yielding parent (C) were genetically marked and crossed in a variety of pairwise combinations, both within and between each group. The results indicated that nuclear fusion had occurred at a very low rate, preventing parasexual analysis of the genetics of cephalosporin production. To overcome the barrier to recombination, protoplasts were prepared and fused. When a group-A strain that synthesized cephalosporin C in high yield from inorganic sulfur but grew slowly and did not sporulate was crossed with a fast-growing, sporulating, low-antibiotic producer from group B, selection of prototrophic progeny gave a strain that appeared to be an unstable heterozygous diploid. This yielded a variety of segregants, some of which exhibited a combination of high yield, fast growth rate, and good sporulation. Since most of the segregants were stable haploids, protoplast fusion appears to be a promising means of using recombination to improve the productivity of *A. chrysogenum*.[285]

VIII. CLONING SYSTEMS

A. Plasmid Vectors

Several plasmids isolated from streptomycetes have been developed as vectors; others are under investigation.[147,151,232] One of the earliest used as a vehicle for transferring DNA between actinomycetes was SCP2*, the variant form of SCP2 that confers increased fertility on its host.[31] This low copy number plasmid from *S. coelicolor* A3(2) has a size of 31 kb; it is indistinguishable from SCP2 in this respect and in the restriction enzyme map obtained with *Pst*I, *Eco*RI, *Bam*HI, and *Hind*III. The map data indicated that a number of sites were available for DNA insertions[33] and the smaller of two fragments obtained by cleavage with *PstI* was replaced with donor DNA carrying the methylenomycin resistance gene to give pSCP111, which retained the ability to replicate in host species, promoted recombination, and was transferred with expression of lethal zygosis. Donor DNA was obtained by digesting the total DNA from SCP$^+$ strains of *S. coelicolor* A3(2) or *S. lividans* 66, and SCP1$^-$ strains of these organisms were used as the initial hosts.[33] The host range of the vector included *S. parvulus* ATCC12434, which is also a conjugational host of SCP2.[62] Other more useful derivatives of SCP2* with deletions and different resistance markers have been prepared and a vector capable of replication in *E. coli* as well as in *Streptomyces* has been constructed by ligation with the *E. coli* plasmid derivative, pBR322.[232]

The SLP1.1 plasmid from *S. lividans* (see Section V.C.2) has been extensively modified for use as a vector. This plasmid, which is smaller (12.5 kb) than SCP2*, but is also maintained at a low copy number (4 to 5 copies per genome), was also used initially to clone the gene for methylenomycin resistance in SCP1$^-$ host strains of *S. coelicolor* A3(2) and *S. lividans* 66 by replacing a DNA segment between two *Pst*I sites with random components of *Pst*I-cleaved DNA from an SCP1$^+$ donor strain.[33] Subsequently, DNA from *S. fradiae* conferring resistance to neomycin (*aph* gene) and DNA from *S. azureus* conferring resistance to thiostrepton (*tsr* gene) were also introduced by shotgun cloning into sites on the plasmid.[34] By replacing nonessential fragments with both the *aph* and *tsr* genes, insertion of foreign DNA at unique *Bam*HI and *Pst*I sites within the *aph* gene could be detected by insertional inactivation. One such vector, pIJ41, was used in the initial stages of cloning the *mel* gene of *S. antibioticus* in *S. lividans* 66,[175] and for cloning into *S. lividans* HH21 a 2-kb fragment of *S. coelicolor* A3(2) DNA containing the *afsB* gene that seems to control formation of A-factor, actinorhodin, and undecylprodigiosin.[153] Although the host range of SLP1-derived vectors is reported to be narrow,[62] *S. reticuli* and *S. venezuelae* have been successfully transformed[392] by the *mel*-containing pIJ41 derivative, pIJ701. Derivates of SLP1.2 have also been constructed to serve as shuttle vectors; pIJ2 carrying the *aph* gene for neomycin resistance was opened at its single *Bcl*I site and ligated with *Bam*HI-cleaved pBR322. The product, pIJ28, replicated stably in its alternative hosts.[62]

The relatively small (8.9 kb) high copy number (40 to 300 copies per genome) plasmid, pIJ101, present in *S. lividans* ISP5434 has a wide host range that includes *S. pristinaespirales, S. albus, S. griseus, S. acrimycini, S. azureus, S. fradiae, S. glaucescens, S. rimosus,* and *S. parvulus.*[177] From it a nonconjugative 4.2-kb cloning vector, pIJ350, has been constructed by replacing the segment between two *Bcl*I sites with DNA carrying the *tsr* gene from *S. azureus*. By inserting from pIJ701 a 1.55-kb fragment containing the *mel* gene of *S. antibioticus,* pIJ350 has been developed further to give the 5.7-kb vector, pIJ702.[175] Several restriction enzyme sites are available within the *mel* gene for inserting foreign DNA, the presence of which can be detected by the absence of brown pigmentation around colonies plated on tyrosine-supplemented media. In an alternative construction, pIJ350 has been developed into a bifunctional vector by inserting DNA composed of pBR322 and a Tn5 fragment carrying the *aph* II gene for kanamycin resistance.[232]

Vectors have been developed from pFJ103, a 20-kb plasmid of the narasin producer, *S. granuloruber,* which has three *Bam*HI restriction sites.[293] Partial digestion with *Bam*HI, ligation to a *Bam*HI fragment carrying *tsr* from *S. azureus,* and transformation of *S. ambofaciens* with the ligated products gave two size classes of recombinant plasmids conferring thiostrepton resistance. The larger, pFJ104 (21.5 kb), in which *tsr* had been inserted at a single *Bam*HI site, replicated in *S. fradiae* and *S. aureofaciens* as well as in *S. ambofaciens.* The smaller, pFJ105 (4.7 kb) contained only a 2.9 -kb fragment of pFJ103 carrying *tsr* and the replication genes. It was modified by introducing DNA from plasmid pBJ322 carrying the *aph* gene which had been inserted into it for subcloning. The resultant bifunctional vector, pFJ123, was stably replicated in both hosts.

S. espinosus contains a small (9.2 kb) cryptic plasmid, pUC6, which has been investigated for its potential as a cloning vector.[225] The copy number, already at about 40 per genome for the native plasmid, was increased 10- to 20-fold by deleting a 2.0-kb *Xho*I fragment to give pUC1061, which retains genes that determine incompatibility with the parent plasmid and contains a number of single restriction enzyme sites potentially available for inserting DNA.[224] Construction of bifunctional plasmids by cloning pUC6 into the *Bam*HI site of pBR322 and pBR325 gave cointegrate derivatives that replicated in *E. coli* but were unstable, about 90% of the progeny losing the plasmids after four subcultures (see Section VI.A).

Deletion mutants of pUC13, the 76-kb plasmid that functions as a prophage of the 84-kb circularly permuted terminally redundant actinophage SF1 of *S. fradiae* (see Section VI.A) have also been examined for potential use as cloning vectors.[65] By digesting the circular prophage genome with *Hind*III and transfecting *S. fradiae* with the ligated products, regions essential for phage replication and maintenance, and those responsible for phage resistance and immunity, were identified. Approximately 10 to 11 kb of DNA in the prophage could be deleted without affecting phage viability. With more extensive deletions, or deletions in specific areas, only the plasmid form was maintained.[65]

A group of small plasmids isolated from various antibiotic-producing streptomycetes[357] and screened for stable maintenance and pock-forming ability in *S. lividans* 66,[244] has been used to clone antibiotic resistance genes of actinomycete orgin.[245] Among the recombinant plasmids, pMS18 carrying a DNA insert for resistance to kanamycin appears promising as a cloning vector. Its host range includes not only *S. lividans* 66 and the neomycin-producing *S. fradiae* SF765 from which it was initially isolated, but also *S. wedmorensis, S. chartreusis* SF1623, *S. griseus, S. ribosidificus,* and *S. hygroscopicus* SF619. A series of conjugative plasmids has recently been isolated from *S. roseochromogenes* and pSRC1b, a high copy number 10.2-kb member of the series, has been developed as a cloning vector by inserting the *tsr* gene for thiostrepton resistance.[334] In addition, the 5.9-kb *S. lavendulae* plasmid, pTA4001 which, like pSRC1b, has the useful property of forming pocks during plasmid transfer, has been modified by introducing DNA containing the gene for streptothricin resistance to form a potential cloning vector.[186]

B. Phage Vectors

The actinophage 0C31 has a moderately broad host range (12 of 20 strains tested)[62] and contains about 40 kb of nonpermuted linear double-stranded DNA with cohesive ends.[338] Suitable deletion mutants into which pBR322 has been inserted provide bifunctional vectors that replicate stably as temperate phage in *Streptomyces* and as plasmids in *E. coli*.[348] The presence of single restriction sites in the pBR322 component provides one opportunity for cloning foreign DNA. Selection for deletion mutants in the chimeric molecule has shown that an 8-kb region of the phage is not essential for plaque formation and thus is also available for cloning. This has allowed construction of vectors such as 0C31 KC400 in which the *attP* site for phage integration into the chromosome has been deleted and the *vph* gene for viomycin resistance from *S. vinaceus* has been introduced.[122] Mature infectious virions are not formed if the DNA packaged is less than 91.5% of the wild-type phage genome size.[208] Thus, when a 0C31::pBR322 vector from which a nonessential *Bam*HI-*Pst*I fragment had been deleted was used to shotgun clone *S. coelicolor* A3(2) DNA cleaved with *Bam*HI and *Pst*I to give fragments in the 2-8 kb size range, the resultant plaques yielded phage that all contained appropriately sized *S. coelicolor* DNA inserts.[62]

C. Host Strains

The most versatile host for cloning DNA from actinomycetes is *S. lividans* 66, although *S. coelicolor* A3(2) has also been used. *S. ambofaciens*,[293] *S. chartreusis* SF1623,[261] and *S. kasugaensis* M518[269] are also reported to lack restriction systems and are, therefore, potentially valuable hosts. Other requirements for a good DNA recipient strain have been outlined.[232,269]

Class II restriction enzymes have been shown to be present in a number of actinomycetes,[298] sometimes in large amounts.[61,94] In a survey of different species, Shimotsu et al.[333] detected activity against *E. coli* phage T4 and λ DNA in only 11 of 55 cell extracts tested. Of these, six appeared to recognize the same CTCGAG sequence. However, restriction systems may be overlooked by biochemical procedures, particularly if the high GC content of actinomycete DNA is not allowed for; differences in the efficiency with which phage infect alternative hosts indicate that restriction and modification phenomena are present, even when specific endonucleases cannot be convincingly demonstrated.[343,383] Other mechanisms may operate but the details remain obscure.[58,244,358,382]

D. Vector Uptake

Conditions suitable for the uptake of phage genome DNA by *Streptomyces* protoplasts were first described by Okanishi et al.[268] A more general procedure was later developed using plasmid SCP2* with *S. coelicolor* A3(2) and *S. parvulus* as the recipients. This was based on the preparation of protoplasts by previously established methods,[152] exposure to covalently closed circular DNA in the presence of polyethylene glycol (PEG 1000), and regeneration of the mycelial form on a high-osmotic medium. Transformation frequencies were above 0.1% for *S. parvulus* and up to 20% for *S. coelicolor*. The procedure was used to introduce hybrid SLP1.2 vector plasmids containing nutritional and resistance genes into *S. lividans* 66[354] and has been adopted, with modifications, for use with other streptomycetes[63,224,359] and for transfections with phage genome DNA.[59,168,190] The methodology has been reviewed by Chater et al.[62]

Although transformants were obtained when linearized SCP2* and SLP1.2 were substituted for the covalently closed circular forms of plasmid DNA, the yields were 10 to 100 times lower.[33] The transformation procedure does not yield recombinants when linear fragments of chromosomal DNA are used. It has been reported[221] that such fragments will transform genetically marked strains at high frequency when entrapped in liposomes, but re-examination with a 0C31 system failed to give an effective increase in transfection efficiency.[301] However, the addition of small DNA-free liposomes to the system did stimulate transfection with phage DNA or transformation with plasmid (though not chromosomal) DNA.

E. Expression of Genes

No difficulties have been encountered in the expression of streptomycete genes after interspecific transfer. Thus, antibiotic resistance genes excised from *S. azureus, S. fradiae,* and *S. vinaceus* have been successfully cloned in *S. lividans* 66 and in *S. chartreusis* SF1623.[245,352-354] Likewise, the *mel* gene from *S. antibioticus* in plasmid pIJ702 causes melanin formation when introduced into a variety of streptomycetes.[175] However, no streptomycete promoter has yet been clearly shown to function as a transcription signal for RNA polymerase in *E. coli.*[155]

When transformed with pBR322 into which the *S. griseus* gene for *p*-aminobenzoic acid synthase *(pab)* had been inserted, *E. coli* mutants with defective *pabA* or *pabB* genes did not initially grow. On prolonged incubation, prototrophic colonies appeared but in these the 4.5-kb *pab*-containing DNA fragment of pBR322 had suffered a 1-kb deletion removing the native *Streptomyces* promoter region. That the *pab* gene had then come under control of the adjacent pBR322 promoter for the tetracycline resistance *(tet)* gene was confirmed by inserting a Tn*5* transposon into the *tet* promoter, thereby silencing both the *tet* and *pab* genes.[232] Readthrough from a host promotor also explains transcription of the *aph* gene from *S. fradiae* in *E. coli*[300,321] where this gene is not normally expressed,[62] and probably of the endoglycosidase H gene from *S. plicatus* as well,[297] although no direct evidence is available in the latter case.

Promoter-probe plasmid vectors for streptomycete genes have been constructed by excising the chloramphenicol resistance *(cat)* gene from *E. coli* plasmid pACYC184 with *Sau*3A and inserting it into the *S. lividans* plasmid SLP1.2.[30] The excised fragment has a region complementary to the 3′-end of 16S *r*RNA in *E. coli* and thus can be translated but lacks an upstream promoter sequence. Expression depends upon the insertion of DNA carrying a promoter into a *Bam*HI site adjacent to the *cat* gene in the vector. Positive results were obtained in *S. lividans* 66 with *Bam*HI digests of DNA from various sources, including not only streptomycetes but also *E. coli, B. licheniformis,* and *Serratia marcescens.*[29] Information on heterologous gene expression has also been obtained by synthesizing a chimeric plasmid from pACYC177 or pACYC184 of *E. coli* and an SLP1.2 derivative of *Streptomyces coelicolor* A3(2). With this, Schottel et al.[310] obtained expression of the *E. coli* genes for chloramphenicol *(cat)* and kanamycin (*aph* III) resistance in *S. lividans,* confirming that ribosomes of *S. lividans* can recognize the Shine and Dalgarno sequence of *E. coli* mRNA, and that RNA polymerase in the streptomycete can transcribe from promoters or nearby promoter-like sequences of the Gram-negative bacterium.

F. Cloning of Genes for Antibiotic Production

A complete gene sequence coding for actinorhodin production has recently been cloned;[223] in addition, progress has been made in cloning specific genes of several other biosynthetic pathways.

S. griseus IMRU3570 contains a *pab* gene specific for *p*-aminobenzoic acid synthase of the candicidin biosynthetic pathway.[105] In one cloning procedure, DNA from a sulfonamide-resistant *S. griseus* mutant that overproduced *p*-aminobenzoic acid was digested with *Bam*HI and the fragments were cloned in pIJ41. The desired *S. lividans* transformants were detected by insertional inactivation of the *aph* gene and screened for resistance to both sulfonamide and thiostrepton. The strain selected contained a plasmid with a 4.5-kb insert specifying *p*-aminobenzoic acid synthesis.[104] An alternative selection procedure made use of an *S. lividans* strain that required *p*-aminobenzoic acid for growth. After transformation as before, neomycin-sensitive, thiostrepton-resistant colonies that grew without a *p*-aminobenzoic acid supplement were isolated and a strain that again carried pIJ41 with a 4.5-kb insert was obtained. Hybridization with this DNA fragment as a radioactive probe demonstrated that the *pab* gene is present only in certain strains of *S. griseus* and *S. fradiae* and absent from species of *Streptomyces* that do not produce candicidin. Presumably, these have an alternative gene for synthesis of *p*-aminobenzoic acid to satisfy nutritional requirements.

Genes from *S. marcescens* encoding some of the enzymes for the biosynthesis of prodigiosin have been cloned in *E. coli* using the cosmid vector, pHC79.[66] Ampicillin-resistant transformants were selected and screened for the presence of prodigiosin pathway genes by testing for cosynthesis of the pigment with blocked mutants. Out of 879 tested, one complemented a mutant that accumulated 4-methoxy-2,2′-bipyrrole-5-carboxaldehyde (MBC) and presumably contained genes for making the other required intermediate, 2-methyl-3-amylpyrrole (MAP); another produced the pigment only when cultured with strains making MBC and MAP, suggesting that it contained the enzyme joining these intermediates. None of the transformants carried genes for the entire pathway of prodigiosin synthesis.

In *S. coelicolor* A3(2), mutants defective in the biosynthesis of undecylprodigiosin were used as recipients for a pIJ702 vector into which DNA fragments of wild-type *S. coelicolor* A3(2) had been inserted.[86] Selection for colonies in which production of the easily detected red-pigmented antibiotic had been restored yielded a strain in which pIJ702 carried a 4.7-kb insert. Introduction of this plasmid complemented two *redE* mutants lacking undecylnorprodigiosin *O*-methyltransferase, the penultimate enzyme in the pathway for biosynthesis of the antibiotic. By labeling the fragment with radioactivity and using it as a hybridization probe to search an *S. coelicolor* A3(2) gene library cloned in *E. coli,* two additional DNA components were isolated, a 9.7-kb *Sst*I fragment complementing *redA, redB,* and *redE* mutants.[147] The success of this technique confirms mapping evidence that the genes are clustered on the chromosome.

Fragments of DNA derived from the sequence of genes coding for methylenomycin biosynthesis in *S. coelicolor* A3(2) have been cloned by using the phage vector 0C31 KC400 (see Section VIII.B).[60] This *attP*-deleted derivative confers viomycin resistance on the recipient only if the phage carries a DNA region homologous to one in the host genome and can thereby integrate via recombination. When the homologous phage DNA is from a gene specifying antibiotic production, integration of the phage can inactivate the corresponding host gene and prevent antibiotic production. Because plasmid SCP1 carrying the genes for methylenomycin synthesis cannot be isolated, donor DNA had to be obtained from the total DNA of an $SCP1^+$ strain. For the donor, *S. parvulus* ATCC12434 was chosen, while an $SCP1^+$ strain of *S. lividans* 66 was used as the host. Since the chromosomal DNAs of these species bear little homology, common sequences were largely confined to SCP1. Detection of lysogens that had acquired viomycin resistance but lost methylenomycin production yielded nine clones in which the phage contained DNA fragments ranging from 2.1 to 2.85 kb in size.[60,147] Three different types of fragment could be distinguished, indicating that at least 7 kb of the coding sequences for methylenomycin biosynthesis had been recovered. Although the cloned segments each contained incomplete genes, they could be used as radioactive probes for hybridization analysis and for isolating the full gene sequences from a cloned genome library. Southern hybridization revealed that the three cloned segments are close to one another and to the methylenomycin resistance gene on SCP1.

Malpartida and Hopwood[223] have succeeded in cloning the complete gene sequence for actinorhodin production. With the plasmid SCP2*-derived vector, pIJ922, 15- to 30-kb fragments of DNA from an act^+ strain of *S. coelicolor* were initially introduced into a class V *act* mutant carrying i *red* mutation to avoid interference from undecylprodigiosin. Of the approximately 8000 transformants selected for thiostrepton resistance and pigment production, one contained a plasmid, pIJ2300, that complemented all *act* classes except VI; a second carried plasmid pIJ2301, complementing *act* classes V, VI, and VII. The 34- and

16-kb DNA inserts within the plasmids contained partially overlapping sequences and a composite 59-kb plasmid, constructed from pIJ2300 and a 4.5-kb segment in pIJ2301 that complemented class VI *act* mtuants was cloned in *S. parvulus* ATCC12434, which does not naturally produce actinorhodin. Selection of transformants gave strains that produced the antibiotic.

IX. CONCLUDING REMARKS

Once considered to be a difficult and unrewarding field where the lack of essential information about producing organisms hampered progress, the genetics of antibiotic production is now amenable to profitable study and is moving into the limelight. Much of the credit for this belongs to the pioneers who have, over the years, laid the foundations for progress by providing an understanding of the genetic systems in the important groups of antibiotic producers. On this base, the new techniques of molecular genetics are being used to build a satisfying explanation of the mechanisms that determine the transfer of information in these organisms and control the formation of secondary metabolites.

With the cloning of DNA carrying genes for antibiotic production, analysis of gene organization and regulation becomes possible and many intriguing aspects of antibiotic production, hitherto explored only at the physiological level, are open to study. A particularly good example is the instability of antibiotic-producing microorganisms which has been a longstanding concern of the fermentation industry. Answers to many questions are now within reach. Most of the recent advances, and many of those now imminent, carry with them a prospect of commercial application. Not only is the way open to more efficient methods of producing biological agents, but the possibility of making new, biosynthetically tailored compounds is at hand.

REFERENCES

1. **Aguilar, A. and Hopwood, D. A.,** Determination of methylenomycin A synthesis by the pSV1 plasmid from *Streptomyces violaceus-ruber* SANK95570, *J. Gen. Microbiol.*, 128, 1893, 1982.
2. **Ahmed, Z. U. and Vining, L. C.,** Evidence for a chromosomal location of the genes coding for chloramphenicol production in *Streptomyces venezuelae*, *J. Bacteriol.*, 154, 239, 1983.
3. **Akagawa, H., Okanishi, M., and Umezawa, H.,** A plasmid involved in chloramphenicol production in *Streptomyces venezuelae:* evidence from genetic mapping, *J. Gen. Microbiol.*, 90, 336, 1975.
4. **Akagawa, H., Okanishi, M., and Umezawa, H.,** Genetics and biochemical studies of chloramphenicol-nonproducing mutants of *Streptomyces venezuelae* carrying plasmid, *J. Antibiot.*, 32, 610, 1979.
5. **Alačević, M.,** Recent advances in *Streptomyces rimosus* genetics, in *Second International Symposium on the Genetics of Industrial Microorganisms,* Macdonald, K. D., Ed., Academic Press, London, 1976, 513.
6. **Alačević, M., Strasek-Vešligaj, M., and Sermonti, G.,** The circular linkage map of *Streptomyces rimosus*, *J. Gen. Microbiol.*, 77, 173, 1973.
7. **Albers-Schonberg, G., Arison, B. H., Kaczka, E., Kahan, F. M., Kahan, J. S., Lago, B., Maiese, W. M., Rhodes, R. E., and Smith, J. L.,** Thienamycin, a new β-lactam antibiotic. III. Structure determination and biosynthetic data, Sixteenth Interscience Conference on Antimicrobial Agents and Chemotherapy, Chicago, Abstr. 229, 1976.
8. **Alikhanian, S. I.,** Induced mutagenesis in the selection of microorganisms, *Adv. Appl. Microbiol.*, 4, 1, 1962.
9. **Alikhanian, S. I. and Borisova, L. N.,** Recombination in *Actinomyces aureofaciens*, *J. Gen. Microbiol.*, 26, 19, 1961.
10. **Alikhanian, S. I., Ilyina, T. S., and Lomovskaya, N. D.,** Transduction in actinomycetes, *Nature (London)*, 188, 245, 1960.
11. **Alikhanian, S. I., Mindlin, S. Z., Goldat, S. U., and Vladimizov, A. V.,** Genetics of organisms producing tetracyclines, *Ann. N.Y. Acad. Sci.*, 81, 914, 1959.

12. **Alikhanian, S. I., Orlova, N. V., Mindlin, S. Z., and Zaitzeva, Z. M.,** Genetic control of oxytetracycline biosynthesis, *Nature (London),* 189, 4768, 1961.
13. **Arnstein, H. R. V., Morris, D., and Toms, E. J.,** Isolation of a tripeptide containing α-aminoadipic acid from the mycelium of *Penicillium chrysogenum* and its possible significance in penicillin biosynthesis, *Biochim. Biophs. Acta,* 35, 561, 1959.
14. **Azevedo, J. L. and Roper, J. A.,** Lethal mutations and balanced lethal systems in *Aspergillus nidulans, J. Gen. Microbiol.,* 49, 149, 1967.
15. **Ball, C.,** Haploidization analysis in *Penicillium chryosogenum, J. Gen. Microbiol.,* 66, 63, 1971.
16. **Ball, C.,** The genetics of *Penicillium chrysogenum, Progr. Ind. Microbiol.,* 12, 47, 1973.
17. **Ball, C.,** Improvement of penicillin productivity in *Penicillium chrysogenum* by recombination, in *First International Symposium on the Genetics of Industrial Microorganisms,* Vaněk, Z., Hošťálek, Z., and Cudlin, J., Eds., Academia, Prague, 1973, 227.
18. **Baltz, R. H.,** Genetic recombination in *Streptomyces fradiae* by protoplast fusion and cell regeneration, *J. Gen. Microbiol.,* 107, 93, 1978.
19. **Baltz, R. H.,** Genetic recombination by protoplast fusion in *Streptomyces, Dev. Ind. Microbiol.,* 21, 43, 1980.
20. **Baltz, R. H.,** Genetics and biochemistry of tylosin production: a model for genetic engineering in antibiotic-producing *Streptomyces,* in *Genetic Engineering of Microorganisms for Chemicals,* Hollaender, A., DeMoss, R. D., Kaplan, S., Konisky, J., Savage, D., and Wolfe, R. S., Eds., Plenum Press, New York, 1982, 431.
21. **Baltz, R. H. and Matsushima, P.,** Protoplast fusion in *Streptomyces.* Conditions for efficient genetic recombination and cell regeneration, *J. Gen. Microbiol.,* 127, 137, 1981.
22. **Baltz, R. H., Seno, E. T., Stonesifer, J., Matsusima, P., and Wild, G. M.,** Genetics and biochemistry of tylosin production by *Streptomyces fradiae,* in *Microbiology — 1981,* Slessinger, D., Ed., American Society for Microbiology, Washington, D.C., 1981, 371.
23. **Baltz, R. H., Seno, E. T., Stonesifer, J., Matsushima, P., and Wild, G. M.,** Genetics and biochemistry of tylosin production, in *Trends in Antibiotic Research. Genetics, Biosyntheses, Actions and New Substances,* Umezawa, H., Demain, A. L., Hata, T., and Hutchinson, C. R., Eds., Japan Antibiotics Research Association, Tokyo, 1982, 65.
24. **Baumann, R., Hütter, R., and Hopwood, D. A.,** Genetic analysis in a melanin-producing streptomycete, *Streptomyces glaucescens, J. Gen. Microbiol.,* 81, 463, 1974.
25. **Beck, J., Gerlach, H., Prelog, V., and Voser, W.,** Stoffwechselprodukte von Actinomyceten. 35 Mitteilung. Über die Konstitution der Makrotetrolide Monactin, Dinactin und Trinactin, *Helv. Chim. Acta,* 45, 620, 1962.
26. **Běhal, V., Bučko, M., and Hošťálek, Z.,** Tetracyclines, in *Biochemistry and Genetic Regulation of Commercially Important Antibiotics,* Vining, L. C., Ed., Addison-Wesley, Reading, Mass., 1983, 255.
27. **Berdy, J., Ed.,** *Handbook of Antibiotic Compounds,* Vols. I to VII, CRC Press, Boca Raton, 1980—1981.
28. **Bibb, M. J.,** *Streptomyces* plasmids: their properties and use as cloning vehicles in antibiotic-producing bacteria, *Microbiology — 1981,* Slessinger, D., Ed., American Society for Microbiology, Washington, D.C., 1981, 367.
29. **Bibb, M. J. and Cohen, S. N.,** Studies on gene expression in *Streptomyces,* in *Genetics of Industrial Microorganisms,* Ikeda, Y. and Beppu, T., Eds., Kodansha, Tokyo, 1982, 49.
30. **Bibb, M. J. and Cohen, S. N.,** Gene expression in *Streptomyces.* Construction and application of promoter-probe plasmid vectors in *Streptomyces lividans, Mol. Gen. Genet.,* 187, 265, 1982.
31. **Bibb, M. J. and Hopwood, D. A.,** Genetic studies of the fertility plasmid SCP2 and its SCP2* variants in *Streptomyces coelicolor* A3(2), *J. Gen. Microbiol.,* 126, 427, 1981.
32. **Bibb, M. J., Freeman, R. F., and Hopwood, D. A.,** Physical and genetical characterization of a second sex factor, SCP2, for *Streptomyces coelicolor* A3(2), *Mol. Gen. Genet.,* 154, 155, 1977.
33. **Bibb, M. J., Schottel, J. L., and Cohen, S. N.,** A DNA cloning system for interspecies gene transfer in antibiotic-producing *Streptomyces, Nature (London),* 284, 526, 1980.
34. **Bibb, M. J., Ward, J. M., and Hopwood, D. A.,** Development of a cloning system for *Streptomyces, Dev. Ind. Microbiol.,* 21, 55, 1980.
35. **Bibb, M. J., Ward, J. M., and Hopwood, D. A.,** Transformation of plasmid DNA into *Streptomyces* at high frequency, *Nature (London),* 274, 398, 1978.
36. **Bibb, M. J., Ward, J. M., Kieser, T., Cohen, S. N., and Hopwood, D. A.,** Excision of chromosomal DNA sequences from *Streptomyces coelicolor* forms a novel family of plasmids detectable in *Streptomyces lividans, Mol. Gen. Genet.,* 184, 230, 1981.
37. **Biswas, G. D. and Sen, S. P.,** Transformation in *Streptomyces* with respect to antibiotic production, *J. Appl. Bacteriol.,* 34, 287, 1971.
38. **Blumauerová, M., Hošťálek, Z., and Vaněk, Z.,** Biosynthesis of tetracyclines: problems and perspectives in genetic analysis, in *Fermentation Technology Today,* Terui, G., Ed., Osaka Society of Fermentation Technology, Osaka, Japan, 1972, 273.

39. **Bonner, D.,** Studies on the biosynthesis of penicillin, *Arch. Biochem.*, 13, 1, 1947.
40. **Boronin, A. M. and Mindlin, S. Z.,** Genetic analysis of *Actinomyces rimosus* mutants with impaired synthesis of antibiotic (in Russian), *Genetika*, 7, 125, 1971.
41. **Boronin, A. M. and Sadovnikova, L. G.,** Elimination by acridine dyes of oxytetracycline resistance in *Actinomyces rimosus*, *Genetika*, 8, 174, 1972.
42. **Bradley, S. G.,** Biosynthesis of an actinorhodin-like pigment by *Streptomyces violaceoruber*, *Dev. Ind. Microbiol.*, 3, 362, 1962.
43. **Bradley, S. G. and Lederberg, J.,** Heterokaryosis in *Streptomyces*, *J. Bacteriol.*, 72, 219, 1956.
44. **Braendle, D. H. and Szybalski, W.,** Genetic interaction among streptomycetes: heterokaryosis and synkaryosis, *Proc. Natl. Acad. Sci. U.S.A.*, 43, 947, 1957.
45. **Braendle, D. H. and Szybalski, W.,** Heterokaryotic compatibility, metabolic cooperation and genic recombination in *Streptomyces*, *Ann. N.Y. Acad. Sci.*, 81, 824, 1959.
46. **Brandsch, R., Hefco, E., Brandsch, C., Rotinberg, P., and Keleman, S.,** Possible plasmid control of sporulation, soluble pigment production and tetracycline resistance in a wild *Streptomyces* sp. strain, *J. Antibiot.*, 33, 1204, 1980.
47. **Brazhnikova, M. G., Lomakina, N. M., Sztaricskai, F., Puskás, M., Makleit, S., and Bognor, R.,** On the antibiotics of vancomycin type, *Kemiai Közlemények*, 27, 143, 1967.
48. **Brockmann, H. and Pini, H.,** Actinorhodin, ein roter Farbstoffe aus Actinomyceten, *Naturwissenschaften*, 34, 190, 1947.
49. **Brownell, G. H.,** Plasmid-like inheritance of EC prophage from matings of lysogenic nocardiae, *Genetics*, 74, 32, 1973.
50. **Bu'Lock, J. D.,** Polyacetylenes and related compounds in nature, in *Progress in Organic Chemistry*, Vol. 6, Cook, J. and Carruthers, W., Eds., Butterworths, London, 1964, 86.
51. **Bu'Lock, J. D.,** Aspects of secondary metabolism in fungi, in *Biogenesis of Antibiotic Substances*, Vaněk, Z. and Hošťálek, Z., Eds., Academic Press, New York, 1965, 61.
52. **Bu'Lock, J. D., Detroy, R. W., Hošťálek, Z., and Munim-Al-Shakarchi, A.,** Regulation of secondary biosynthesis in *Gibberella fujijuroi*, *Trans. Br. Mycol. Soc.*, 62, 377, 1974.
53. **Byng, G. S. and Turner, J. M.,** Incorporation of (^{14}C)shikimate into phenazines and their further metabolism by *Pseudomonas phenazinium*, *Biochem. J.*, 164, 139, 1977.
54. **Calam, C. T., Daglish, L. B., and Gaitskell, W. S.,** Hybridization experiments with *Penicillium patulum* and *Fusarium moniliforme*, in *Genetics of Industrial Microorganisms: Actinomycetes and Fungi*, Vaněk, Z., Hošťálek, Z., and Cudlin, J., Eds., Academia, Prague, 1973, 265.
55. **Calam, C. T., Daglish, L. B., and McCann, E. P.,** Penicillin: tactics in strain improvement, in *Second International Symposium on the Genetics of Industrial Microorganisms*, Macdonald, K. D., Ed., Academic Press, London, 1976, 273.
56. **Chakrabarty, S. L.,** Ultraviolet-induced mutations in *Streptomyces indicus*, *Mutat. Res.*, 8, 513, 1969.
57. **Chang, L. T., Behr, D. A., and Elander, R. P.,** The effects of intercalating agents on kanamycin production and other phenotypic characteristics of *Streptomyces kanamycetiucs*, *Dev. Ind. Microbiol.*, 21, 233, 1980.
58. **Chater, K. F.,** Restriction in *Streptomyces*, *Zentralbl. Bakteriol. Parasitenkd. Infectionskr. Hyg. Abt. 1, Suppl.*, 303, 1978.
59. **Chater, K. F.,** Actinophage DNA, *Dev. Ind. Microbiol.*, 21, 65, 1980.
60. **Chater, K. F.,** The deployment of *Streptomyces* vectors, in *Genetics of Industrial Microorganisms*, Ikeda, Y. and Beppu, T., Eds., Kodansha, Tokyo, 1982, 71.
61. **Chater, K. F. and Wilde, L. C.,** Restriction of a bacteriophage of *Streptomyces albus* G involving endonuclease *Sal* I, *J. Bacteriol.*, 128, 644, 1976.
62. **Chater, K. F., Hopwood, D. A., Kieser, T., and Thompson, C. J.,** Gene cloning in *Streptomyces*, *Curr. Top. Microbiol. Immunol.*, 96, 69, 1982.
63. **Chung, S. -T.,** Isolation and characterization of *Streptomyces fradiae* plasmids which are prophage of the actinophage SF1, *Gene*, 17, 239, 1982.
64. **Chung, S.-T. and Molnar, M.,** Characterization and development of *Streptomyces* gene-cloning vectors. I. *Streptomyces fradiae*, in *Trends in Antibiotic Research. Genetics, Biosyntheses, Actions and New Substances*, Umezawa, H., Demain, A. L., Hata, T., and Hutchinson, C. R., Eds., Japan Antibiotics Research Association, Tokyo, 1982, 42.
65. **Coats, J. H. and Roeser, J.,** Genetic recombination in *Streptomyces bikiniensis* var. *zorbonensis*, *J. Bacteriol.*, 105, 880, 1971.
66. **Dauenhauer, S. A., Hull, R. A., and Williams, R. P.,** Cloning and expression in *Escherichia coli of Serratia marcescens* genes encoding prodigiosin biosynthesis, *J. Bacteriol.*, 158, 1128, 1984.
67. **Davies, J. E. and Yagisawa, M.,** The aminocyclitol glycosides (aminoglycosides), in *Biochemistry and Genetic Regulation of Commercially Important Antibiotics*, Vining, L. C., Ed., Addison-Wesley, Reading, Mass., 1983, 329.

68. **Delic, V., Hopwood, D. A., and Friend, E. J.,** Mutagenesis by *N*-methyl-*N'* nitro-*N*-nitrosoguanidine, *Mutat. Res.*, 9, 182, 1970.
69. **Demain, A. L.,** How do antibiotic-producing microorganisms avoid suicide?, *Ann. N.Y. Acad. Sci.*, 235, 601, 1974.
70. **Demain, A. L., Kupka, J., Shen, Y.-Q., and Wolfe, S.,** Microbiological synthesis of β-lactam antibiotics, in *Trends in Antibiotic Research. Genetic, Biosyntheses, Actions and New Substances,* Umezawa, H., Demain, A. L., Hata, T., and Hutchinson, C. R., Eds., Japan Antibiotics Research Association, Tokyo, 1982, 233.
71. **Demain, A. L., Aharonowitz, Y., and Martin, J.-F.,** Metabolic control of secondary biosynthetic pathways, in *Biochemistry and Genetic Regulation of Commercially Important Antibiotics,* Vining, L. C., Ed., Addison-Wesley, Reading, Mass., 1983, 49.
72. **Dimroth, P., Walter, H., and Lynen, F.,** Biosynthese von 6-Methylsalicylsäure, *Eur. J. Biochem.*, 13, 98, 1970.
73. **Ditchburn, P., Giddings, B., and Macdonald, K. D.,** Rapid screening for the isolation of mutants of *Aspergillus nidulans* with increased penicillin yields, *J. Appl. Bacteriol.*, 37, 515, 1974.
74. **Ditchburn, P., Holt, G., and Macdonald, K. D.,** The genetic location of mutations increasing penicillin yield in *Aspergillus nidulans,* in *Second International Symposium on the Genetics of Industrial Microorganisms,* Macdonald, K. D., Eds., Academic Press, London, 1976, 213.
75. **Dobritsa, A. P., Dobritsa, S. V., and Tanyashin, V. I.,** Isolation and characterization of plasmid from the *Bacillus brevis* var. G. B. cells, *Mol. Gen. Genet.*, 164, 195, 1978.
76. **Dobrzanski, W. T. and Osowiecki, H.,** Isolation and some properties of the competence factor from group H *Streptococcus* strain CHALLIS, *J. Gen. Microbiol.*, 48, 299, 1967.
77. **Doull, J. L., Vining, L. C., and Stuttard, C.,** A cryptic plasmid in the chloramphenicol-producing actinomycete, *Streptomyces phaeochromogenes, FEMS Microbiol. Lett.*, 16, 349, 1983.
78. **Dowding, J. E.,** Characterization of a bacteriophage virulent for *Streptomyces coelicolor* A3(2), *J. Gen. Microbiol.*, 76, 163, 1973.
79. **Drew, S. W. and Demain, A. L.,** Stimulation of cephalosporin production by methionine peptides in a mutant blocked in reverse transsulfuration, *J. Antibiot.*, 28, 889, 1975.
80. **Dulaney, E. L. and Dulaney, D. D.,** Mutant populations of *Streptomyces viridifaciens, Trans. N.Y. Acad. Sci.*, 29, 782, 1967.
81. **Edwards, G. F. St. L., Holt, G., and Macdonald, K. D.,** Mutants of *Aspergillus nidulans* impaired in penicillin biosynthesis, *J. Gen. Microbiol.*, 84, 420, 1974.
82. **Elander, R. P.,** Trends in the genetic improvement of antibiotic-producing microorganisms, in *Trends in Antibiotic Research. Genetics, Biosyntheses, Actions and New Substances,* Umezawa, H., Demain, A. L., Hata, T., and Hutchinson, C. R., Eds., Japan Antibiotics Research Association, Tokyo, 1982, 16.
83. **Elander, R. P., Espenshade, M. A., Pathak, S. G., and Pan, C. H.,** The use of parasexual genetics in an industrial strain improvement program with *Penicillium chrysogenum,* in *First International Symposium on the Genetics of Industrial Microorganisms,* Vaněk, Z., Hošťálek, Z., and Cudlin, J., Eds., Academia, Prague, 1973, 239.
84. **El-Kresh, T. and Vezina, C.,** Plasmid determination of antimycin production in auxotrophs of *Streptomyces* sp. M-506, *Abstr. Third Int. Symp. Genetics of Industrial Microorganisms,* 1978, 5.
85. **Elson, S. W., Oliver, R. S., Bycroft, B. W., and Faruk, E. A.,** Studies on the biosynthesis of clavulanic acid. III. Incorporation of DL-[3,4,-$^{13}C_2$]glutamic acid, *J. Antibiot.*, 35, 81, 1982.
86. **Feitelson, J. S. and Hopwood, D. A.,** Cloning of a gene for an *O*-methyl-transferase involved in antibiotic biosynthesis, *Mol. Gen. Genet.*, 190, 394, 1983.
87. **Fodor, K. and Alfoldi, L.,** Fusion of protoplasts in *Bacillus megaterium, Proc. Natl. Acad. Sci. U.S.A.*, 73, 2147, 1976.
88. **Freeman, R. F., Bibb, M. J., and Hopwood, D. A.,** Chloramphenicol acetyltransferase-independent chloramphenicol resistance in *Streptomyces coelicolor* A3(2), *J. Gen. Microbiol.*, 98, 453, 1977.
89. **Friebel, T. E. and Demain, A. L.,** Oxygen-dependent inactivation of gramicidin S synthetase in *Bacillus brevis, J. Bacteriol.*, 130, 1010, 1977.
90. **Friedrich, C. G. and Demain, A. L.,** Homocitrate synthase as the crucial site of the lysine effect on penicillin biosynthesis, *J. Antibiot.*, 30, 760, 1977.
91. **Friend, E. J. and Hopwood, D. A.,** The linkage map of *Streptomyces rimosus, J. Gen. Microbiol.*, 68, 187, 1971.
92. **Friend, E. J., Warren, M., and Hopwood, D. A.,** Genetic evidence for a plasmid controlling fertility in an industrial strain of *Streptomyces rimosus, J. Gen. Microbiol.*, 106, 201, 1978.
93. **Froyshov, O., Zimmer, T. L., and Laland, S. G.,** Biosynthesis of microbial peptides by the thiotemplate mechanism, in *International Review of Biochemistry. Amino Acid and Protein Biosynthesis II,* Vol. 18, Arnstein, H. R. V., Ed., University Park Press, Baltimore, Md., 1978, 49.
94. **Fuchs, L. Y., Covarrubias, L., Escalante, L., Sanchez, S., and Bolivar, F.,** Characterization of a site-specific restriction endonuclease *Sph*I from *Streptomyces phaeochromogenes, Gene,* 10, 39, 1980.

95. **Fuchs, P. G., Zajdel, J., and Dobrzanski, W. T.,** Possible plasmid nature of the determinant for production of the antibiotic nisin in some strains of *Streptococcus lactis, J. Gen. Microbiol.*, 88, 189, 1975.
96. **Fujisawa, Y., Shirafuji, H., Kida, M., Nara, K., Yoneda, M., and Kanzaki, T.,** Accumulation of deacetylcephalosporin C by cephalosporin-C negative mutants of *Cephalosporium acremonium, Agric. Biol. Chem.*, 39, 1295, 1975.
97. **Fjuisawa, Y., Kitano, K., and Kanzaki, T.,** Accumulation of deacetoxycephalosporin C by a deacetylcephalosporin mutant of *Cephalosporium acremonium, Agric. Biol. Chem.*, 39, 2049, 1975.
98. **Furumai, T., Takeda, K., and Okanishi, M.,** Function of plasmids in the production of aureothricin. I. Elimination of plasmids and alteration of phenotypes caused by protoplast regeneration in *Streptomyces kasugaensis, J. Antibiot.*, 35, 1367, 1982.
99. **Gallo, M. and Katz, E.,** Regulation of secondary metabolite biosynthesis. Catabolite repression of phenoxazinone synthetase and actinomycin formation by glucose, *J. Bacteriol.*, 109, 659, 1972.
100. **Gatenbeck, S. and Sierankiewicz, J.,** On the biosynthesis of tenuazonic acid in *Alternaria tenuis, Acta Chem. Scand.*, 27, 1825, 1973.
101. **Gaucher, G. M., Lam, K. S., GrootWassink, J. W. D., Neway, J., and Deo, Y. M.,** The initiation and longevity of patulin biosynthesis, *Dev. Ind. Microbiol.*, 22, 219, 1981.
102. **Gerber, N. N., McInnes, A. G., Smith, D. G., Walter, J. A., Wright, J. L. C., and Vining, L. C.,** Biosynthesis of prodiginines. ^{13}C resonance assignments and enrichment patterns in nonyl-, cyclononyl-, methylcyclodecyl- and butylcycloheptyl-prodiginines produced by actinomycete cultures supplemented with ^{13}C-labeled acetate and ^{15}N-labeled nitrate, *Can. J. Chem.*, 56, 1155, 1978.
103. **Ghisalba, O., Auden, J. A. L., Schupp, T., and Neusch, J.,** The rifamycins: properties, biosynthesis and fermentation, in *Biotechnology of Industrial Antibiotics,* Vandamme, E. J., Ed., Marcel Dekker, New York, 1984, 281.
104. **Gil, J. A. and Hopwood, D. A.,** Cloning and expression of a *p*-aminobenzoic acid synthetase gene of the candicidin producer, *Streptomyces griseus, Gene,* 25, 119, 1983.
105. **Gil, J. A., Naharro, G., Villaneuva, J. R., and Martin, J.-F.,** Regulation by aromatic amino acids of the *p*-aminobenzoic acid synthetase level in *S. griseus,* in *Advances in Biotechnology,* Vol. 3, *Fermentation Products,* Vezina, C. and Singh, K., Eds., Pergamon Press, New York, 1981, 141.
106. **Godfrey, O.,** Directed mutations in *Streptomyces lipmanii, Can. J. Microbiol.*, 20, 1479, 1974.
107. **Godfrey, O., Ford, L., and Huber, M. L. B.,** Interspecies matings of *Streptomyces fradiae* with *Streptomyces bikiniensis* mediated by conventional and protoplast fusion techniques, *Can. J. Microbiol.*, 24, 994, 1978.
108. **Gotfredsen, W. O. and Jahnsen, S.,** Fusidane-type antibiotics, *Process Biochem.*, 4, 11, 1969.
109. **Goulden, S. A. and Chattaway, F. W.,** End-product control of acetohydroxy acid synthetase by valine in *Penicillium chrysogenum* Q176 and a high penicillin-yielding mutant, *J. Gen. Microbiol.*, 59, 111, 1969.
110. **Gräfe, U., Reinhardt, G., Krebs, D., Eritt, I., and Steudel, A.,** Effect of the autoregulator from *Streptomyces griseus* JA5142 on surface cultures of blocked mutant ZIMET43682, *Z. Allgemein. Mikrobiol.*, 23, 359, 1983.
111. **Gregory, K. F. and Huang, J. C. C.,** Tyrosinase inheritance in *Streptomyces scabies.* I. Genetic recombination, *J. Bacteriol.*, 87, 1281, 1964.
112. **Gregory, K. F. and Huang, J. C. C.,** Tyrosinase inheritance in *Streptomyces scabies.* II. Induction of tyrosinase deficiency by acridine dyes, *J. Bacteriol.*, 87, 1287, 1964.
113. **Gregory, K. F. and Shyu, W. J.,** Apparent cytoplasmic inheritance of tyrosinase competence in *Streptomyces scabies, Nature (London),* 191, 465, 1961.
114. **Grove, J. F.,** Griseofulvin, in *Antibiotics,* Vol. 2, *Biosynthesis,* Gottlieb, D. and Shaw, P. D., Eds., Springer, New York, 1967, 123.
115. **Guerola, N., Ingraham, J. L., and Cerda-Olmedo, E.,** Induction of closely-linked multiple mutations by nitrosoguanidine, *Nature (London) New Biol.*, 230, 122, 1971.
116. **Hamlyn, P. F. and Ball, C.,** Recombination studies with *Cephalosporium acremonium,* in *Third International Symposium on the Genetics of Industrial Microorganisms,* Sebek, O. K. and Laskin, A. I., Eds., American Society for Microbiology, Washington, D.C., 1979, 185.
117. **Hara, O. and Beppu, T.,** Mutants blocked in streptomycin production in *Streptomyces griseus* — the role of A-factor, *J. Antibiot.*, 35, 349, 1982.
118. **Hara, O. and Beppu, T.,** Induction of streptomycin-inactivating enzyme by A-factor in *Streptomyces griseus, J. Antibiot.*, 35, 1208, 1982.
119. **Hara, O., Horinouchi, S., Uozumi, T., and Beppu, T.,** Genetic analysis of A-factor sysnthesis in *Streptomyces coelicolor* A3(2) and *Streptomyces griseus, J. Gen. Microbiol.*, 129, 2939, 1983.
120. **Harold, R. J. and Hopwood, D. A.,** Ultraviolet-sensitive mutants of *Streptomyces coelicolor.* II. Genetics, *Mutat. Res.*, 10, 439, 1970.
121. **Harold, R. J. and Hopwood, D. A.,** A rapid method for complementation testing of ultraviolet-sensitive *(uvs)* mutants of *Streptomyces coelicolor, Mutat. Res.*, 16, 27, 1972.

122. **Harris, J. E., Chater, K. F., Bruton, C. J., and Piret, J. M.,** The restriction mapping of c gene deletions in *Streptomyces* bacteriophage C31 and their use in cloning vector development, *Gene,* 22, 167, 1983.
123. **Haskins, R. H.,** Biochemistry of the Ustilaginales. I. Preliminary cultural studies of *Ustilago zeae, Can. J. Res.,* C28, 213, 1950.
124. **Hayakawa, T., Tanaka, T., Sakaguchi, K., Otake, N., and Yonehara, H.,** A linear plasmid-like DNA in *Streptomyces* sp. producing lankacidin group antibiotics, *J. Gen. Appl. Microbiol.,* 25, 255, 1979.
125. **Hintermann, G., Crameri, R., Kieser, T., and Hütter, R.,** Restriction analysis of the *Streptomyces glaucescens* genome by agarose gel electrophoresis, *Arch. Microbiol.,* 130, 218, 1981.
126. **Hintermann, G., Crameri, R., Kieser, T., and Hütter, R.,** A technique for the detection of large DNA alterations in complex genomes, *Anal. Biochem.,* 121, 327, 1982.
127. **Hirochika, H. and Sakaguchi, K.,** Analysis of linear plasmids isolated from *Streptomyces:* association of protein with the ends of the plasmid DNA, *Plasmid,* 7, 59, 1982.
128. **Hirota, Y.,** The effect of acridine dyes on mating-type factors in *Escherichia coli, Proc. Natl. Acad. Sci. U.S.A.,* 46, 57, 1960.
129. **Holloway, B. W.,** Genetic organization in *Pseudomonas,* in *Genetics and Biochemistry of Pseudomonas,* Clarke, P. H. and Richmond, M. H., Eds., Wiley, London, 1975, 133.
130. **Holt, G., Edwards, G. F. St. L., and Macdonald, K. D.,** The genetics of mutants impaired in the biosynthesis of penicillin, in *Second International Symposium on the Genetics of Industrial Microorganisms,* Macdonald, K. D., Ed., Academic Press, London, 1976, 199.
131. **Hopwood, D. A.,** New data on the linkage map of *Streptomyces coelicolor, Genetics,* 54, 1174, 1965.
132. **Hopwood, D. A.,** Genetic analysis and genome structure in *Streptomyces coelicolor, Bacteriol. Rev.,* 31, 373, 1967.
133. **Hopwood, D. A.,** Genome topology and mapping in *Streptomyces coelicolor,* in *Genetics and Breeding of Streptomyces,* Sermonti, G. and Alačević, M., Eds., Yugoslav Academy of Sciences and Arts, Zagreb, 1969, 5.
134. **Hopwood, D. A.,** Extrachromosomally determined antibiotic production, *Annu. Rev. Microbiol.,* 32, 373, 1978.
135. **Hopwood, D. A.,** Genetics of antibiotic production by actinomycetes, *J. Nat. Products,* 42, 596, 1979.
136. **Hopwood, D. A.,** Uses of protoplasts in the genetic manipulation of streptomycetes, in *Actinomycetes,* Schaal, K. P. and Pulverer, G., Eds., *Zbl. Bakt.,* Suppl. 11, Gustav Fischer, Stuttgart, 1981, 523.
137. **Hopwood, D. A.,** Actinomycete genetics and antibiotic production, in *Biochemistry and Genetic Regulation of Commercially Important Antibiotics,* Vining, L. C., Ed., Addison-Wesley, Reading, Mass., 1983, 1.
138. **Hopwood, D. A. and Glauert, A. M.,** Observations on the chromatinic bodies of *Streptomyces coelicolor, J. Biophys. Biochem. Cytol.,* 8, 257, 1960.
139. **Hopwood, D. A. and Merrick, M. J.,** Genetics of antibiotic production, *Bacteriol. Rev.,* 41, 595, 1977.
140. **Hopwood, D. A. and Sermonti, G.,** The genetics of *Streptomyces coelicolor, Adv. Genet.,* 11, 273, 1962.
141. **Hopwood, D. A. and Wright, H.,** Genetic transformation in a thermophilic actinomycete, *Heredity,* 27, 483, 1971.
142. **Hopwood, D. A. and Wright, H. M.,** A plasmid of *Streptomyces coelicolor* carrying a chromosomal locus and its interspecific transfer, *J. Gen. Microbiol.,* 79, 331, 1973.
143. **Hopwood, D. A. and Wright, H. M.,** Interactions of the plasmid SCP1 with the chromosome of *Streptomyces coelicolor* A3(2), in *Second International Symposium on the Genetics of Industrial Microorganisms,* Macdonald, K. D., Ed., Academic Press, London, 1976, 607.
144. **Hopwood, D. A. and Wright, H. M.,** Genetic studies on SCP1-prime strains of *Streptomyces coelicolor* A3(2), *J. Gen. Microbiol.,* 95, 107, 1976.
145. **Hopwood, D. A. and Wright, H. M.,** Bacterial protoplast fusion: recombination in fused protoplasts of *Streptomyces coelicolor, Mol. Gen. Genet.,* 162, 307, 1978.
146. **Hopwood, D. A. and Wright, H. M.,** Factors affecting recombinant frequency in protoplast fusions of *Streptomyces coelicolor, J. Gen. Microbiol.,* 111, 137, 1979.
147. **Hopwood, D. A., Bibb, M. J., Bruton, C. J., Chater, K. F., Feitelson, J. F., and Gil, J. A.,** Cloning *Streptomyces* genes for antibiotic production, *Trends Biotechnol.,* 1, 42, 1983.
148. **Hopwood, D. A., Chater, K. F., Dowding, J. E., and Vivian, A.,** Advances in *Streptomyces coelicolor* genetics, *Bacteriol. Rev.,* 37, 371, 1973.
149. **Hopwood, D. A., Harold, R. J., Vivian, A., and Ferguson, H. M.,** A new kind of fertility variant in *Streptomyces coelicolor, Genetics,* 62, 461, 1969.
150. **Hopwood, D. A., Kieser, T., Wright, H. M., and Bibb, M. J.,** Plasmids, recombination and chromosome mapping in *Streptomyces lividans* 66, *J. Gen. Microbiol.,* 129, 2257, 1983.
151. **Hopwood, D. A., Thompson, C. J., Kieser, T., Ward, J. M., and Wright, H. M.,** Progress in the development of plasmid cloning vectors for *Streptomyces,* in *Microbiology — 1981,* Schlessinger, D., Ed., American Society for Microbiology, Washington, D.C., 1981, 376.
152. **Hopwood, D. A., Wright, H. M., Bibb, M. J., and Cohen, S. N.,** Genetic recombination through protoplast fusion in *Streptomyces, Nature (London),* 268, 171, 1977.

153. **Horinouchi, S., Hara, O., and Beppu, T.,** Cloning of a pleiotropic gene that positively controls biosynthesis of A-factor, actinorhodin and prodigiosin in *Streptomyces coelicolor* A3(2) and *Streptomyces lividans*, *J. Bacteriol.*, 155, 1238, 1983.
154. **Horinouchi, S., Kumada, Y., and Beppu, T.,** Unstable genetic determinant of A-factor biosynthesis in streptomycin-producing organisms: cloning and characterization, *J. Bacteriol.*, 158, 481, 1984.
155. **Horinouchi, S., Uozumi, T., and Beppu, T.,** Cloning of *Streptomyces* DNA into *Escherichia coli:* absence of heterospecific gene expression of *Streptomyces* genes in *E. coli*, *Agric. Biol. Chem.*, 44, 367, 1980.
156. **Hornemann, U. and Hopwood, D. A.,** Biosynthesis of methylenomycin A, a plasmid-determined antibiotic, in *Antibiotics*, Vol. 4, *Biosynthesis*, Corcoran, J. W., Ed., Springer, Berlin, 1981, 123.
157. **Hosoda, J., Tani, N., Konomi, T., Ohsawa, S., Aoki, H., and Imanaka, H.,** Incorporation of ^{14}C-amino acids into nocardicin A by growing cells, *Agric. Biol. Chem.*, 41, 2007, 1977.
158. **Hošťálek, Z., Blumauerová, M., and Vaněk, Z.,** Tetracyclines, in *Economic Microbiology*, Vol. 3, *Secondary Products of Metabolism*, Rose, A. H., Ed., Academic Press, London, 1979, 293.
159. **Hotta, K., Okami, Y., and Umezawa, H.,** Elimination of the ability of a kanamycin producing strain to biosynthesize deoxystreptamine moiety by acriflavine, *J. Antibiot.*, 30, 1146, 1978.
160. **Hurst, A.,** Nisin, *Adv. Appl. Microbiol.*, 27, 85, 1981.
161. **Hutchinson, C. R.,** The biosynthesis of tetracyline and anthracycline antibiotics, in *Antibiotics*, Vol. 4 *Biosynthesis*, Corcoran, J. W., Ed., Springer, Berlin, 1981, 1.
162. **Hütter, R., Kieser, T., Crameri, R., and Hintermann, G.,** Chromosomal instability in *Streptomyces glaucescens*, in *Actinomycetes*, Schaal, K. P. and Pulverer, G., Eds., *Zbl. Bakt.* Supp. 11, Gustav Fischer, Stuttgart and New York, 1981, 551.
163. **Ichikawa, T., Date, M., Ishikura, T., and Ozaki, A.,** Improvement of kasugamycin-producing strain by agar-piece method and the prototroph method, *Folia Microbiol. (Prague)*, 16, 218, 1971.
164. **Ikeda, H., Inoue, M., and Ōmura, S.,** Improvement of macrolide antibiotic-producing streptomycete strains by the regeneration of protoplasts, *J. Antibiot.*, 36, 283, 1983.
165. **Ikeda, H., Tanaka, H., and Ōmura, S.,** Isolation and characterization of covalently closed circular DNA associated with chromosomal and membrane fraction from *Streptomyces ambofaciens*, *J. Antibiot.*, 35, 497, 1982.
166. **Ikeda, H., Tanaka, H., and Ōmura, S.,** Genetic and biochemical features of spiramycin biosynthesis in *Streptomyces ambofaciens* — curing, protoplast regeneration and plasmid transfer, *J. Antibiot.*, 35, 507, 1982.
167. **Inouye, S., Tsuruoka, T., Ito, T., and Niida, T.,** Structure and synthesis of nojirimycin, *Tetrahedron*, 24, 2125, 1968.
168. **Isogai, T., Takahashi, H., and Saito, H.,** High-frequency protoplast transfection of *Streptomyces parvulus* 2297 with actinophage R4 DNA, *Agric. Biol. Chem.*, 44, 2425, 1980.
169. **Jacob, F., Brenner, S., and Cuzin, F.,** Regulation of DNA replication in bacteria, *Cold Spring Harbor Symp. Quant. Biol.*, 28, 329, 1963.
170. **Jones, L. A. and Taylor, A. R.,** Incomplete and aberrant particles of actinophage MSP2 from lysogenates of *Streptomyces venezuelae*, *J. Virol.*, 14, 493, 1974.
171. **Jones, A. and Westlake, D. W. S.,** Regulation of chloramphenicol synthesis in *Streptomyces* sp. 3022a. Properties of arylamine synthetase, an enzyme involved in antibiotic biosynthesis, *Can. J. Microbiol.*, 20, 1599, 1974.
172. **Kähler, R. and Noack, D.,** Action of acridine orange and ethidium bromide on growth and antibiotic activity of *Streptomyces hygroscopicus* JA6599, *Z. Allg. Microbiol.*, 14, 529, 1974.
173. **Katz, E. and Demain, A. L.,** The peptide antibiotics of *Bacillus:* biogenesis and possible functions, *Bacteriol. Rev.*, 41, 449, 1977.
174. **Katz, E. and Weissbach, H.,** Incorporation of ^{14}C-labeled amino acids into actinomycin and protein by *Streptomyces antibioticus*, *J. Biol. Chem.*, 238, 666, 1963.
175. **Katz, E., Thompson, C. J., and Hopwood, D. A.,** Cloning and expression of the tyrosinase gene from *Streptomyces antibioticus* in *Streptomyces lividans*, *J. Gen. Microbiol.*, 129, 2703, 1983.
176. **Khokhlov, A. S., Tovarova, I. I., Borisova, L. N., Pliner, S. A., Shevchenko, L. A., Kornitskaya, E. Y., Ivkina, N. S., and Rapaport, I. A.,** A-factor responsible for the biosynthesis of streptomycin by a mutant strain of *Actinomyces streptomycini*, *Doklady Akad. Nauk SSSR*, 177, 232, 1967.
177. **Kieser, T., Hopwood, D. A., Wright, H. M., and Thompson, C. J.,** pIJ101, a multicopy, broad-host-range *Streptomyces* plasmid: functional analysis and development of DNA-cloning vectors, *Mol. Gen. Genet.*, 185, 223, 1982.
178. **Kirby, R.,** Genetic mapping of *Streptomyces clavuligerus*, *FEMS Microbiol. Lett.*, 3, 177, 1978.
179. **Kirby, R.,** An unstable genetic element affecting the production of the antibiotic holomycin by *Streptomyces clavuligerus*, *FEMS Microbiol. Lett.*, 3, 283, 1978.
180. **Kirby, R. and Hopwood, D. A.,** Genetic determination of methylenomycin synthesis by the SCP1 plasmid of *Streptomyces coelicolor* A3(2), *J. Gen. Microbiol.*, 98, 239, 1977.

181. **Kirby, R., Wright, L. F., and Hopwood, D. A.,** Plasmid determined antibiotic synthesis and resistance in *Streptomyces coelicolor, Nature (London),* 254, 265, 1975.
182. **Kleiner, E. M., Pliner, S. A., Soifer, V. S., Onoprienko, V. V., Balashova, T. A., Rozynov, B. V., and Khokhlov, A. S.,** Structure of the A-factor, a bioregulator from *Streptomyces griseus, Bioorg. Khim.,* 2, 1142, 1976.
183. **Kleinkauf, H. and von Döhren, H.,** Nucleic acid-independent synthesis of peptides, *Curr. Top. Microbiol. Immunol.,* 91, 129, 1981.
184. **Kleinkauf, H. and von Döhren, H., Eds.,** *Peptide Antibiotics — Biosynthesis and Functions,* de Gruyter, Berlin, 1982.
185. **Kleinkauf, H. and von Döhren, H.,** Peptides, in *Biochemistry and Genetic Regulation of Commercially Important Antibiotics,* Vining, L. C., Ed., Addison-Wesley, Reading, Mass., 1983, 95.
186. **Kobayashi, T., Shimotsu, H., Horinouchi, S., Uozumi, T., and Beppu, T.,** Isolation and characterization of a pock-forming plasmid, pTA4001, from *Streptomyces lavendulae, J. Antibiot.,* 37, 368, 1984.
187. **Komatsu, K., Leboul, J., Harford, S., and Davies, J.,** Studies of plasmids in neomycin-producing *Streptomyces fradiae,* in *Microbiology — 1981,* Schlessinger, D., Ed., American Society for Microbiology, Washington, D.C., 1981, 384.
188. **Kominek, L. A. and Sebek, O.,** Biosynthesis of novobiocin and related coumarin antibiotics, *Dev. Ind. Microbiol.,* 15, 60, 1974.
189. **Kozak, W., Rajchert-Trzpil, M., and Dobrzanski, W. T.,** The effect of proflavin, ethidium bromide and an elevated temperature on the appearance of nisin-negative clones in nisin-producing strains of *Streptococcus lactis, J. Gen. Microbiol.,* 83, 295, 1974.
190. **Krugel, H., Fiedler, G., and Noack, D.,** Transfection of protoplasts from *Streptomyces lividans* 66 with actinophage SH10 DNA, *Mol. Gen. Genet.,* 177, 297, 1980.
191. **Kurahashi, K.,** Biosynthesis of peptide antibiotics, in *Antibiotics,* Vol. 4, Biosynthesis, Corcoran, J. W., Ed., Springer, Berlin, 1981, 325.
192. **Kützner, H. J. and Waksman, S. A.,** *Streptomyces coelicolor* Muller and *Streptomyces violaceoruber* Waksman and Curtiss, two distinctly different organisms, *J. Bacteriol.,* 78, 528, 1959.
193. **Lago, B. D. and Streicher, S. L.,** New genetics of streptomycetes, *ASM News,* 48, 53, 1982.
194. **Laland, S. G. and Zimmer, T.-L.,** The protein template mechanism of synthesis for the peptide antibiotics produced by *Bacillus brevis, Essays Biochem.,* 9, 31, 1973.
195. **Lancini, G.,** Ansamycins, in *Biochemistry and Genetic Regulation of Commercially Important Antibiotics,* Vining, L. C., Ed., Addison-Wesley, Reading, Mass., 1983, 231.
196. **Lara, F., Mateos, R. C., Vasquez, G., and Sanchez, S.,** Induction of penicillin biosynthesis by L-glutamate in *Penicillium chrysogenum, Biochem. Biophys. Res. Commun.,* 105, 172, 1982.
197. **Laskin, A. I. and Lechevalier, H. A., Eds.,** *Handbook of Microbiology,* Vol. V, *Microbial Products,* CRC Press, Boca Raton, 1984.
198. **Lemke, P. A. and Nash, C. H.,** Mutations that affect antibiotic synthesis by *Cephalosporium acremonium, Can. J. Microbiol.,* 18, 255, 1972.
199. **Lepesant-Kejzlarova, J., Lepesant, J.-A., Walle, J., Billault, A., and Dedonder, R.,** Revision of the linkage map of *Bacillus subtilis* 168: indications for circularity of the chromosome, *J. Bacteriol.,* 121, 823, 1975.
200. **Liersch, M., Nuesch, J., and Treichler, H. J.,** Final steps in the biosynthesis of cephalosporin C, in *Second International Symposium on the Genetics of Industrial Microorganisms,* Macdonald, K. D., Ed., Academic Press, London, 1976, 179.
201. **Lilley, G., Clark, A. E., and Lawrence, G. C.,** Control of the production of cephamycin C and thienamycin by *Streptomyces cattleya* NRRL8057, *J. Chem. Tech. Biotechnol.,* 31, 127, 1981.
202. **Lipmann, F.,** Attempts to map a process evolution of peptide biosynthesis, *Science,* 173, 875, 1971.
203. **Lipmann, F.,** Bacterial production of antibiotic polypeptides by thiol-linked synthesis on protein templates, *Adv. Microb. Physiol.,* 21, 227, 1980.
204. **Liras, P., Villanueva, J. R., and Martin, J.-F.,** Sequential expression of macromolecule biosynthesis and candicidin formation in *Streptomyces griseus, J. Gen. Microbiol.,* 102, 269, 1977.
205. **Liu, C.-M., Hermann, T., and Miller, P. A.,** Feedback inhibition of the synthesis of an antibiotic, aurodox (X-5108), *J. Antibiot.,* 30, 244, 1977.
206. **Liu, C.-M., McDaniel, L. E., and Schaffner, C. P.,** Studies on candicidin biogenesis, *J. Antibiot.,* 25, 116, 1972.
207. **Loder, P. B. and Abraham, E. P.,** Isolation and nature of intracellular peptides from a cephalosporin C-producing *Cephalosporium* sp., *Biochem. J.,* 123, 471, 1971.
208. **Lomovskaya, N. D., Chater, K. F., and Mkrtumian, N. M.,** Genetics and molecular biology of *Streptomyces* bacteriophages, *Bacteriol. Rev.,* 44, 206, 1980.
209. **Lomovskaya, N. D., Mkrtumian, N. M., Gostimskaya, N. L., and Danilenko, V.,** Characterization of temperate actinophage 0C31 isolated from *Streptomyces coelicolor* A3(2), *J. Virol.,* 9, 258, 1972.

210. **Macdonald, K. D.,** Preservation of the heterozygous diploid condition in industrial microorganisms, *Nature (London),* 204, 404, 1964.
211. **Macdonald, K. D.,** The persistence of parental genome segregation in *Penicillium chrysogenum* after nitrogen mustard treatment, *Mutat. Res.,* 5, 302, 1968.
212. **Macdonald, K. D.,** Fungal genetics and antibiotic production, in *Biochemistry and Genetic Regulation of Commercially Important Antibiotics,* Vining, L. C., Ed., Addison-Wesley, Reading, Mass., 1983, 25.
213. **Macdonald, K. D. and Holt, G.,** Genetics of biosynthesis and overproduction of penicillin, *Sci. Prog. (Oxford),* 63, 547, 1976.
214. **Macdonald, K. D., Holt, G., and Ditchburn, P.,** The genetics of penicillin production, in *Proceedings of the Fourth International Fermentation Symposium, Fermentation Technology Today,* Terui, G., Ed., Society of Fermentation Technology, Japan, 1972, 251.
215. **Macdonald, K. D., Hutchinson, J. M., and Gillett, W. A.,** Isolation of auxotrophs of *Penicillium chrysogenum* and their penicillin yields, *J. Gen. Microbiol.,* 33, 365, 1963.
216. **Macdonald, K. D., Hutchinson, J. M., and Gillett, W. A.,** Heterokaryon studies and the genetic control of penicillin and chrysogenin production in *Penicillium chrysogenum, J. Gen. Microbiol.,* 33, 375, 1963.
217. **Macdonald, K. D., Hutchinson, J. M., and Gillett, W. A.,** Formation and segregation of heterozygous diploids between a wild-type strain and derivatives of high penicillin yield in *Penicillium chrysogenum, J. Gen. Microbiol.,* 33, 385, 1963.
218. **Macdonald, K. D., Hutchinson, J. M., and Gillett, W. A.,** Properties of heterozygous diploids between strains of *Penicillium chrysogenum* selected for high penicillin yield, *Antonie van Leeuwenhoek,* 30, 209, 1964.
219. **Macdonald, K. D., Hutchinson, J. M., and Gillett, W. A.,** Heterozygous diploids of *Penicillium chrysogenum* and their segregation pattern, *Genetica,* 36, 378, 1965.
220. **Majumdar, M. K. and Majumdar, S. K.,** Synthesis of neomycin by washed mycelium of *Streptomyces fradiae* and some physiological considerations, *Folia Microbiol.,* 16, 285, 1971.
221. **Makins, J. F. and Holt, G.,** Liposome-mediated transformation of streptomycetes by chromosomal DNA, *Nature (London),* 293, 671, 1981.
222. **Malik, V. S.,** Genetics of applied microbiology, *Adv. Genet.,* 20, 37, 1979.
223. **Malpartida, F. and Hopwood, D. A.,** Molecular cloning of a whole biosynthetic pathway of a *Streptomyces* antibiotic and its expression in a heterologous host, *Nature (London),* 309, 462, 1984.
224. **Manis, J. J. and Highlander, S. K.,** Partial characterization of a small, multicopy plasmid from *Streptomyces espinosus* and the derivation of a high copy-number deletion mutant, *Gene,* 18, 13, 1982.
225. **Manis, J. J. and Highlander, S. K.,** Characterization and development of *Streptomyces* gene-cloning vectors. II. *Streptomyces espinosus,* in *Trends in Antibiotic Research. Genetics, Biosyntheses, Actions and New Substances,* Umezawa, H., Demain, A. L., Hata, T., and Hutchinson, C. R., Eds., Japan Antibiotics Research Association, Tokyo, 1982, 48.
226. **Marahiel, M. A., Danders, W., Krause, M., and Kleinkauf, H.,** Biological role of gramicidin S in spore functions. Studies on gramicidin-S-negative mutants of *Bacillus brevis* ATCC9999, *Eur. J. Biochem.,* 99, 49, 1979.
227. **Marahiel, M. A., Lurz, R., and Kleinkauf, H.,** Characterization of chromosomal and membrane associated plasmid in *Bacillus brevis* ATCC9999, *J. Antibiot.,* 34, 323, 1981.
228. **Martin, J.-F.,** Control of antibiotic synthesis by phosphate, *Adv. Biochem. Eng.,* 6, 105, 1977.
229. **Martin, J.-F.,** Polyene macrolide antibiotics, in *Economic Microbiology,* Vol. 3, *Secondary Products of Metabolism,* Rose, A. H., Ed., Academic Press, London, 1979, 355.
230. **Martin, J.-F.,** Polyenes, in *Biochemistry and Genetic Regulation of Commercially Important Antibiotics,* Vining, L. C., Ed., Addison-Wesley, Reading, Mass., 1983, 207.
231. **Martin, J.-F., and Demain, A. L.,** Control by phosphate of candicidin production, *Biochem. Biophys. Res. Commun.,* 71, 1103, 1976.
232. **Martin, J.-F. and Gil, J. A.,** Cloning and expression of antibiotic production genes, *Biotechnology,* 2, 63, 1984.
233. **Martin, J.-F., Alegre, M. T., Gil, J. A., and Naharro, G.,** Characterization of mutants of *Streptomyces griseus* deregulated or blocked in candicidin biosynthesis: molecular mechanism of control of antibiotic biosynthesis, in *Advances in Biotechnology,* Vol. 3, *Fermentation Products,* Moo-Young, M., Vezina, C., and Singh, K., Eds., Pergamon Press, Toronto, 1981, 129.
234. **Matselyukh, B. P.,** Structure and function of *Actinomyces olivaceus* genome, in *Second International Symposium on the Genetics of Industrial Microorganisms,* Macdonald, K. D., Ed., Academic Press, New York, 1976, 553.
235. **Matsubara-Nakano, M., Kataoka, Y., and Ogawara, H.,** Unstable mutation of β-lactamase production in *Streptomyces lavendulae, Antimicrob. Agents Chemother.,* 17, 124, 1980.
236. **Mattick, A. T. R. and Hirsch, A.,** A powerful inhibitory substance produced by group-N streptococci, *Nature (London),* 154, 551, 1944.

237. **McCormick, J. R. D.,** Tetracyclines, in *Antibiotics,* Vol. 2, *Biosynthesis,* Gottlieb, D. and Shaw, P. D., Eds., Springer, Berlin, 1967, 113.
238. **McCormick, J. R. D.,** Point-blocked mutants and the biogenesis of tetracycline, in *Genetics and Breeding of Streptomyces,* Sermonti, G. and Alačević, M., Eds., Yugoslav Academy of Sciences and Arts, Zagreb, 1969, 163.
239. **Mendelovitz, S. and Aharonowitz, Y.,** Regulation of cephamycin C synthesis, aspartokinase, dehydrodipicolinic acid synthetase and homoserine dehydrogenase by aspartic acid family amino acids in *Streptomyces clavuligerus, Antimicrob. Agents Chemother.,* 21, 74, 1982.
240. **Mendelovitz, S. and Aharonowitz, Y.,** β-Lactam antibiotic production by *Streptomyces clavuligerus* mutants impaired in regulation of aspartokinase, *J. Gen. Microbiol.,* 129, 2063, 1983.
241. **Micheli, M., Sermonti, G., and Lanfaloni, L.,** Evidence of heterokaryosis in *Streptomyces coelicolor* A3(2), *Mol. Gen. Genet.,* 194, 541, 1984.
242. **Michelson, A. M. and Vining, L. C.,** Loss of chloramphenicol production in strains of *Streptomyces* species 3022a treated with acriflavine and ethidium bromide, *Can. J. Microbiol.,* 24, 662, 1978.
243. **Mindlin, S. Z., Zayceva, Z. M., Germanov, A. B., and Shishkina, T. A.,** The genetic analysis of "inactive" mutants of *Actinomyces rimosus,* in *Antibiotics: Advances in Research, Production and Clinical Use,* Herold, M. and Gabriel, Z., Eds., Butterworth, London, 1966, 695.
244. **Murakami, T., Nojiri, C., Toyama, H., Hayashi, E., Yamada, Y., and Nagaoka, K.,** Pock-forming plasmids from antibiotic-producing *Streptomyces, J. Antibiot.,* 36, 429, 1983.
245. **Murakami, T., Nojiri, C., Toyama, H., Hayashi, E., Katumata, K., Anzai, H., Matsuhashi, Y., Yamada, Y., and Nagaoka, K.,** Cloning of antibiotic resistance genes in *Streptomyces, J. Antibiot.,* 36, 1305, 1983.
246. **Nakano, M. M., Ishihara, H., and Ogawara, H.,** Fusion of protoplasts of *Streptomyces lavendulae, J. Antibiot.,* 35, 359, 1982.
247. **Nakano, M. M., Ozawa, K., and Ogawara, H.,** Possible involvement of a plasmid in arginine auxotrophic mutation of *Streptomyces kasugaensis, J. Bacteriol.,* 3, 1501, 1980.
248. **Nash, C. H., de la Higuera, N., Neuss, N., and Lemke, P. A.,** Application of biochemical genetics to the biosynthesis of β-lactam antibiotics, *Dev. Ind. Microbiol.,* 15, 114, 1974.
249. **Neft, N. and Farley, T. M.,** Studies on the biosynthesis of antimycin A. I. Incorporation of ^{14}C-labeled metabolites into the 3-formamidosalicylyl moiety, *J. Antibiot.,* 25, 298, 1972.
250. **Nelson, T. C.,** Screening methods in microbiology, *Mutation and Plant Breeding,* National Research Council Publication No. 891, National Academy of Sciences, Washington, D.C., 1961, 311.
251. **Neway, J. and Gaucher, G. M.,** Intrinsic limitations on the continued production of the antibiotic patulin by *Penicillium urticae, Can. J. Microbiol.,* 27, 206, 1981.
252. **Nishio, C., Komura, S., and Kurahashi, K.,** Peptide antibiotic subtilin is synthesized via precursor proteins, *Biochem. Biophys. Res. Commun.,* 116, 751, 1983.
253. **Nojiri, C., Watabe, H., Katsumata, K., Yamada, Y., Murakami, T., and Kumata, Y.,** Isolation and characterization of plasmids from parent and variant strains of *Streptomyces ribosidificus, J. Antibiot.,* 33, 118, 1980.
254. **Normansell, P. J., Normansell, I. D., and Holt, G.,** Genetic and biochemical studies of mutants of *Penicillium chrysogenum* impaired in penicillin production, *J. Gen. Microbiol.,* 112, 113, 1979.
255. **Novick, R., Gruss, A., and Edelman, I.,** Plasmid curing during the formation and regeneration of protoplasts in *Staphylococcus aureus,* in *Plasmids and Transposons. Environmental Effects and Maintenance Mechanisms,* Stuttard, C. and Rozee, K. R., Eds., Academic Press, Toronto, 1980, 303.
256. **Ochi, K. and Katz, E.,** The possible involvement of plasmid(s) in actinomycin synthesis by *Streptomyces parvulus* and *Streptomyces antibioticus, J. Antibiot.,* 31, 1143, 1978.
257. **Ochi, K. and Katz, E.,** Genetic analysis of the actinomycin-producing determinants (plasmid) in *Streptomyces parvulus* using protoplast fusion technique, *Can. J. Microbiol.,* 26, 1460, 1980.
258. **Ochi, K., Hitchcock, M. J. M., and Katz, E.,** High-frequency fusion of *Streptomyces parvulus* or *Streptomyces antibioticus* protoplasts induced by polyethylene glycol, *J. Bacteriol.,* 139, 984, 1979.
259. **Ogata, S., Suenaga, H., and Hayashida, S.,** Pock formation of *Streptomyces endus* with production of phage taillike particles, *Appl. Environ. Microbiol.,* 43, 1182, 1982.
260. **Ogata, S., Yoshino, S., Suenaga, H., Aoyama, K., Kitajima, N., and Hayashida, S.,** Specific lysogenicity in *Streptomyces azureus, Appl. Environ. Microbiol.,* 42, 135, 1981.
261. **Ogawara, H.,** Antibiotic resistance in pathogenic and producing bacteria, with special reference to β-lactam antibiotics, *Microbiol. Rev.,* 45, 591, 1981.
262. **Ogawara, H. and Nozaki, S.,** Effect of acriflavin on the production of β-lactamase in *Streptomyces, J. Antibiot.,* 30, 337, 1977.
263. **Okanishi, M.,** Plasmids and antibiotic synthesis in streptomycetes, in *Genetics of Industrial Microorganisms,* Sebek, O. K. and Laskin, A. I., Eds., American Society for Microbiology, Washington, D.C., 1979, 134.

264. **Okanishi, M.,** Plasmid involvement in antibiotic production, in *Molecular Breeding and Genetics of Applied Microorganisms,* Sakaguchi, K. and Okanishi, M., Eds., Academic Press, New York, 1980, 29.
265. **Okanishi, M.,** Function of plasmids in aureothricin production, in *Trends in Antibiotic Research. Genetics, Biosyntheses, Actions and New Substances,* Umezawa, H., Demain, A. L., Hata, T., and Hutchinson, C. R., Eds., Japan Antibiotics Research Association, Tokyo, 1982, 32.
266. **Okanishi, M. and Okami, Y.,** Isolation and characterization of an actinophage active on *Streptomyces kanamyceticus, J. Gen. Microbiol.,* 12, 207, 1966.
267. **Okanishi, M. and Umezawa, H.,** Plasmids involved in antibiotic production in *Streptomyces,* in *Genetics of Actinomycetales,* Freerksen, E., Tarnok, I., and Thumin, J. H., Eds., Gustav Fischer, Stuttgart, 1978, 19.
268. **Okanishi, M., Hamana, K., and Umezawa, H.,** Factors affecting infection of protoplasts with deoxyribonucleic acid of actinophage PK-66, *J. Virol.,* 2, 686, 1968.
269. **Okanishi, M., Katagiri, K., Furumai, T., Takeda, K., Kawaguchi, K., Satoh, M., and Nabeshima, S.,** Basic techniques for DNA cloning and conditions requried for streptomycetes as a host, *J. Antibiot.,* 36, 99, 1983.
270. **Okanishi, M., Manome, T., and Umezawa, H.,** Isolation and characterization of plasmid DNAs in actinomycetes, *J. Antibiot.,* 33, 88, 1980.
271. **Okanishi, M., Ohta, T., and Umezawa, H.,** Possible control of formation of aerial mycelium and antibiotic production in *Streptomyces* by episomic factors, *J. Antibiot.,* 23, 45, 1970.
272. **Okanishi, M., Suzuki, K., and Umezawa, H.,** Formation and regeneration of streptomycete protoplasts: cultural conditions and morphological study, *J. Gen. Microbiol.,* 80, 389, 1974.
273. **Okazaki, T., Kitahara, T., and Okami, Y.,** Studies on marine microorganisms. IV. A new antibiotic SS-228Y produced by *Chainia* isolated from shallow sea mud, *J. Antibiot.,* 28, 176, 1975.
274. **Ōmura, S. and Nakagawa, A.,** Biosynthesis of 16-membered macrolide antibiotics, in *Antibiotics,* Vol. 4, *Biosynthesis,* Corcoran, J. W., Ed., Springer, Berlin, 1981, 175.
275. **Ōmura, S. and Tanaka, Y.,** Macrolides, in *Biochemistry and Genetic Regulation of Commercially Important Antibiotics,* Vining, L. C., Ed., Addison-Wesley, Reading, Mass., 1983, 179.
276. **Ōmura, S., Ikeda, H., and Kitao, C.,** Detection of a plasmid in *Streptomyces ambofaciens* KA-1028 and its possible involvement in spiramycin production, *J. Antibiot.,* 32, 1058, 1979.
277. **Ōmura, S., Ikeda, H., and Tanaka, H.,** Extraction and characterization of plasmids from macrolide antibiotic-producing streptomycetes, *J. Antibiot.,* 34, 478, 1981.
278. **Ōmura, S., Katagiri, M., Nakagawa, A., Sano, Y., Nomura, S., and Hata, T.,** Studies on cerulenin. V. Structure of cerulenin, *J. Antibiot.,* Ser. A, 20, 349, 1967.
279. **Onishi, N., Izaki, K., and Takahashi, H.,** A macrocyclic antibiotic M-230B produced by *Myxococcus xanthus.* Isolation and characterization, *J. Antibiot.,* 37, 13, 1984.
280. **Ono, H., Hintermann, G., Crameri, R., and Hütter, R.,** Reiterated sequences in a mutant strain of *Streptomyces glaucescens* and cloning of the sequence in *Escherichia coli, Mol. Gen. Genet.,* 186, 106, 1982.
281. **O'Sullivan, J. and Ball, C.,** β-Lactams, in *Biochemistry and Genetic Regulation of Commercially Important Antibiotics,* Vining, L. C., Ed., Addison-Wesley, Reading, Mass., 1983, 73.
282. **O'Sullivan, J., Gillum, A. M., Aklonis, C. A., Souser, M. L., and Sykes, R. B.,** Biosynthesis of monobactam compounds. Origin of the carbon atoms of the β-lactam ring, *Antimicrob. Agents Chemother.,* 21, 558, 1982.
283. **Pearce, C. J. and Rinehart, K. L., Jr.,** Biosynthesis of aminocyclitol antibiotics, in *Antibiotics,* Vol. 4, *Biosynthesis,* Corcoran, J. W., Ed., Springer, Berlin, 1981, 74.
284. **Peberdy, J. F.,** Fungal protoplasts: isolation, reversion and fusion, *Annu. Rev. Microbiol.,* 33, 21, 1979.
285. **Peberdy, J. F.,** New approaches to gene transfer in fungi, in *Third International Symposium on the Genetics of Industrial Microorganisms,* Sebek, O. K. and Laskin, A. I., Eds., American Society for Microbiology, Washington, D.C., 1979, 192.
286. **Pigac, J., Vesligaj, M., and Delic, V.,** Mapping of oxytetracycline genes in *Streptomyces rimosus,* in *Genetics of Industrial Microorganisms, Abstr. 2nd. Int. Symp.,* Academic Press, London, 1974, 105.
287. **Pontecorvo, G., Roper, J. A., Hemmons, L. M., Macdonald, K. D., and Bufton, A. W. J.,** The genetics of *Aspergillus nidulans, Adv. Genet.,* 5, 141, 1953.
288. **Queener, S. W., Capone, J. J., Radue, A. B., and Nagarajan, R.,** Synthesis of deacetoxycephalosporin C by a mutant of *Cephalosporium acremonium, Antimicrob. Agents Chemother.,* 6, 334, 1974.
289. **Randazzo, R., Sciandrello, G., Carere, R., Bignami, M., Velcich, A., and Sermonti, G.,** Localized mutagenesis in *Streptomyces coelicolor, Mutat. Res.,* 36, 291, 1976.
290. **Randazzo, R., Sermonti, G., Carere, R., and Bignami, M.,** Comutation in *Streptomyces, J. Bacteriol.,* 113, 500, 1973.
291. **Redshaw, P. A., McCann, P. A., Sankaran, L., and Pogell, B. M.,** Control of differentiation in streptomycetes: involvement of extrachromosomal deoxyribonucleic acid and glucose repression in aerial mycelia development, *J. Bacteriol.,* 125, 698, 1976.

292. **Redshaw, P. A., McCann, P. A., Pentella, M. A., and Pogell, B. M.,** Simultaneous loss of multiple differentiated functions in aerial mycelium negative isolates of *Streptomycetes, J. Bacteriol.*, 137, 891, 1979.
293. **Richardson, M. A., Mabe, J. A., Beerman, N. E., Nakatsukasa, W. M., and Fayerman, J. T.,** Development of cloning vehicles from the *Streptomyces* plasmid pFJ103, *Gene,* 20, 451, 1982.
294. **Rinehart, K. L., Jr.,** Biosynthesis and mutasynthesis of aminocyclitol antibiotics, *J. Antibiot.*, 32, S32, 1979.
295. **Rinehart, K. L., Jr., Potgeiter, M., Jin, W.-Z., Pearce, C. J., Wright, D. A., Wright, J. L. C., Walter, J. A., and McInnes, A. G.,** Biosynthetic studies on antibiotics employing stable isotopes, in *Trends in Antibiotic Research. Genetics, Biosyntheses, Actions and New Substances,* Umezawa, H., Demain, A. L., Hata, T., and Hutchinson, C. R., Eds., Japan Antibiotics Research Association, Tokyo, 1982, 171.
296. **Rhodes, P. M., Winskill, N., Friend, E. J., and Warren, M. W.,** Biochemical and genetic characterization of *Streptomyces rimosus* mutants impaired in oxytetracycline biosynthesis, *J. Gen. Microbiol.*, 124, 329, 1981.
297. **Robbins, P. W., Wirth, D. F., and Hering, C.,** Expression of the *Streptomyces* enzyme endoglycosidase H in *Escherichia coli, J. Biol. Chem.*, 256, 10640, 1981.
298. **Roberts, R. J.,** Restriction and modification enzymes and their recognition sequences, *Gene,* 8, 329, 1980.
299. **Robinson, M., Lewis, E., and Napier, E.,** Occurrence of reiterated DNA sequences in strains of *Streptomyces* produced by an interspecific protoplast fusion, *Mol. Gen. Genet.*, 182, 336, 1981.
300. **Rodgers, W. H., Springer, W., and Young, F. E.,** Cloning and expression of a *Streptomyces fradiae* neomycin-resistance gene in *Escherichia coli, Gene,* 18, 133, 1982.
301. **Rodicio, M.-R. and Chater, K. F.,** Small DNA-free liposomes stimulate transfection of *Streptomyces* protoplasts, *J. Bacteriol.*, 151, 1079, 1982.
302. **Rodicio, M.-R., Manzanal, M.-B., and Hardisson, C.,** Protoplast formation during spore germination in *Streptomyces, Curr. Microbiol.*, 1, 89, 1978.
303. **Rudd, B. A. M. and Hopwood, D. A.,** Genetics of actinorhodin biosynthesis by *Streptomyces coelicolor* A3(2), *J. Gen. Microbiol.*, 114, 35, 1979.
304. **Rudd, B. A. M. and Hopwood, D. A.,** A pigmented mycelial antibiotic in *Streptomyces coelicolor:* control by a chromosomal gene cluster, *J. Gen. Microbiol.*, 119, 333, 1980.
305. **Ryter, A. and Landman, O. E.,** Electron microscope study of the relationship between mesosome loss and the stable L state (or protoplast state) in *Bacillus subtilis, J. Bacteriol.*, 88, 457, 1964.
306. **Sagara, Y., Fukui, K., Ota, F., Yoshida, N., Kashiyama, T., and Fujimoto, M.,** Rapid formation of protoplasts of *Streptomyces griseoflavus* and their fine structure, *Jpn. J. Microbiol.*, 15, 73, 1971.
307. **Saito, H. and Ikeda, Y.,** An improved method for isolating biochemical mutants of *Streptomyces griseoflavus, J. Gen. Appl. Microbiol.*, 3, 240, 1957.
308. **Sankaran, L. and Pogell, B. M.,** Biosynthesis of puromycin in *Streptomyces alboniger:* regulation and properties of *O*-demethylpuromycin *O*-methyltransferase, *Antimicrob. Agents Chemother.*, 8, 721, 1975.
309. **Schaeffer, P., Cami, B., and Hotchkiss, R. D.,** Fusion of bacterial protoplasts, *Proc. Natl. Acad. Sci. U.S.A.*, 73, 2151, 1976.
310. **Schottel, J. L., Bibb, M. J., and Cohen, S. L.,** Cloning and expression in *Streptomyces lividans* of antibiotic resistance genes derived from *Escherichia coli, J. Bacteriol.*, 146, 360, 1981.
311. **Schrempf, H.,** Functions of plasmids in producers of macrolides, in *Advances in Biotechnology* Vol. 3, *Fermentation Products,* Vezina, C. and Singh, K., Eds., Pergamon Press, Willowdale, Ont., 1980, 25.
312. **Schrempf, H.,** Functions of plasmid genes in streptomycetes, in *Actinomycetes,* Schaal, K. P. and Pulverer, G., Eds., *Zbl. Bakt.*, Suppl. 11, Gustav Fischer, Stuttgart, 1981, 545.
313. **Schrempf, H.,** Plasmid loss and changes within the chromosomal DNA of *Streptomyces reticuli, J. Bacteriol.*, 151, 701, 1982.
314. **Schrempf, H.,** Reiterated sequences within the genome of *Streptomyces,* in *Genome Rearrangement,* Chater, K. F., Cullis, C. A., Hopwood, D. A., Johnson, A. W. B., and Woolhouse, H. W., Eds., Croom Helm, London, 1983, 131.
315. **Schrempf, H.,** Deletion and amplification of DNA sequences in melanin-negative variants of *Streptomyces reticuli, Mol. Gen. Genet.*, 189, 501, 1983.
316. **Schrempf, H. and Goebel, W.,** Characterization of a plasmid from *Streptomyces coelicolor* A3(2), *J. Bacteriol.*, 131, 251, 1977.
317. **Schrempf, H. and Goebel, W.,** Functions of plasmid genes in *Streptomyces reticuli,* in *Plasmids of Medical, Environmental and Commercial Importance,* Timmis, K. N. and Pühler, A., Eds., Elsevier/North-Holland, Amsterdam, 1979, 259.
318. **Schrempf, H., Bujard, H., Hopwood, D. A., and Goebel, W.,** Isolation of covalently-closed circular deoxyribonucleic acid from *Streptomyces coelicolor* A3(2), *J. Bacteriol.*, 121, 416, 1975.
319. **Schupp, T. and Nuesch, J.,** Chromosomal mutations in the final step of rifamycin B biosynthesis, *FEMS Microbiol. Lett.*, 6, 23, 1979.

320. **Schupp, T., Hütter, R., and Hopwood, D. A.,** Genetic recombination in *Nocardia mediterranei*, *J. Bacteriol.*, 121, 128, 1975.
321. **Schupp, T., Toupet, C., Stahlhammar-Carlemalm, M., and Meyer, J.,** Expression of a neomycin phosphotransferase gene from *Streptomyces fradiae* in *Escherichia coli* after interplasmidic recombination, *Mol. Gen. Genet.*, 189, 27, 1983.
322. **Sermonti, G.,** Genetics of penicillin production, *Ann. N.Y. Acad. Sci.*, 81, 950, 1959.
323. **Sermonti, G.,** *Genetics of Antibiotic-Producing Microorganisms*, Wiley-Interscience, London, 1969.
324. **Sermonti, G. and Casciano, S.,** Sexual polarity in *Streptomyces coelicolor*, *J. Gen. Microbiol.*, 33, 293, 1963.
325. **Sermonti, G. and Lanfaloni, L.,** Antibiotic genes — their assemblage and localization in *Streptomyces*, in *Overproduction of Microbial Products*, Krumphanzl, V., Sikyta, B., and Vaněk, Z., Eds., Academic Press, London, 1982, 487.
326. **Sermonti, G., Bandiera, M., and Spada-Sermonti, I.,** New approach to the genetics of *Streptomyces coelicolor*, *J. Bacteriol.*, 91, 384, 1966.
327. **Sermonti, G., Lanfaloni, L., and Micheli, M. R.,** A jumping gene in *Streptomyces coelicolor* A3(2), *Mol. Gen. Genet.*, 177, 453, 1980.
328. **Sermonti, G., Lanfaloni, L., and Micheli, M. R.,** Properties of transposon SCTn*1* of *Streptomyces coelicolor* A3(2), *Mol. Gen. Genet.*, 191, 158, 1983.
329. **Sermonti, G., Mancinelli, A., and Spada-Sermonti, I.,** Heterogeneous clones (heteroclones) in *Streptomyces coelicolor* A3(2), *Genetics*, 45, 669, 1960.
330. **Sermonti, G., Petris, A., Micheli, M., and Lanfaloni, L.,** A factor involved in chloramphenicol resistance in *Streptomyces coelicolor* A3(2): its transfer in the absence of fertility factor, *J. Gen. Microbiol.*, 100, 347, 1977.
331. **Sermonti, G., Petris, A., Micheli, M., and Lanfaloni, L.,** Chloramphenicol resistance in *Streptomyces coelicolor* A3(2) — possible involvement of a transposable element, *Mol. Gen. Genet.*, 164, 99, 1978.
332. **Shaw, P. D. and Piwowarsky, J.,** Effects of ethidium bromide and acriflavine on streptomycin production by *Streptomyces bikiniensis*, *J. Antibiot.*, 30, 404, 1977.
333. **Shimotsu, H., Takahashi, H., and Saito, H.,** Site-specific endonucleases in *Streptomyces* strains, *Agric. Biol. Chem.*, 44, 1665, 1980.
334. **Shindoh, Y., Nakano, M. M., and Ogawara, H.,** Pock-forming plasmids isolated from *Streptomyces roseochromogenes*, *J. Antibiot.*, 37, 512, 1984.
335. **Shirahama, T., Furmai, T., and Okanishi, M.,** A modified regeneration method for streptomycete protoplasts, *Agric. Biol. Chem.*, 45, 1271, 1981.
336. **Simpson, I. N. and Caten, C. E.,** Selection for increased penicillin titre following hybridization of divergent lines of *Aspergillus nidulans*, *J. Gen. Microbiol.*, 126, 311, 1981.
337. **Skurray, R. A. and Reeves, P.,** Characterization of lethal zygosis associated with conjugation in *Escherichia coli K12*, *J. Bacteriol.*, 113, 58, 1973.
338. **Sladkova, I. A., Lomovskaya, N. D., and Chinenova, T. A.,** The structure and size of the genome of actinophage 0C31 of *Streptomyces coelicolor* A3(2), *Genetika*, 13, 342, 1977.
339. **Snoke, J. E.,** Formation of bacitracin by protoplasts of *Bacillus licheniformis*, *J. Bacteriol.*, 81, 986, 1961.
340. **Somerson, N. L., Demain, A. L., and Nunheimer, T. D.,** Reversal of lysine inhibition of penicillin production by α-aminoadipic acid, *Arch. Biochem. Biophys.*, 93, 238, 1961.
341. **Spizizen, J.,** Transformation of biochemically deficient strains of *Bacillus subtilis* by deoxyribonucleate, *Proc. Natl. Acad. Sci. U.S.A.*, 44, 1072, 1958.
342. **Stuttard, C.,** Transduction of auxotrophic markers in a chloramphenicol-producing strain of *Streptomyces*, *J. Gen. Microbiol.*, 110, 479, 1979.
343. **Stuttard, C.,** Temperate phages of *Streptomyces venezuelae:* lysogeny and host specificity shown by phages SV1 and SV2, *J. Gen. Microbiol.*, 128, 115, 1982.
344. **Stuttard, C.,** Localized hydroxylamine mutagenesis and cotransduction of threonine and lysine genes in *Streptomyces venezuelae*, *J. Bacteriol.*, 155, 1219, 1983.
345. **Stuttard, C.,** Cotransduction of *his* and *trp* loci in *Streptomyces venezuelae*, *FEMS Micobiol. Lett.*, in press.
346. **Su, T. L.,** Antibiotic-producing organisms in faeces and sewage, *Br. J. Exp. Pathol.*, 29, 466, 1948.
347. **Suarez, J. E. and Chater, K. F.,** Polyethylene glycol-assisted transfection of *Streptomyces protoplasts*, *J. Bacteriol.*, 142, 8, 1980.
348. **Suarez, J. E. and Chater, K. F.,** DNA cloning in *Streptomyces:* a bifunctional replicon comprising pBR322 inserted into a *Streptomyces* phage, *Nature (London)*, 286, 527, 1980.
349. **Suhadolnik, R. J.,** Biosynthesis of the nucleoside antibiotics, in *Antibiotics*, Vol. 4, *Biosynthesis*, Corcoran, J. W., Ed., Springer, Berlin, 1981, 353.

350. **Sykes, R. B., Parker, W. L., and Wells, J. S.,** β-Lactam antibiotics produced by bacteria, in *Trends in Antibiotic Reserach. Genetics, Biosyntheses, Actions and New Substances,* Umezawa, H., Demain, A. L., Hata, T., and Hutchinson, C. R., Eds., Japan Antibiotic Research Association, Tokyo, 1982, 115.
351. **Takeda, K., Kawaguchi, K., and Okanishi, M.,** Extraction, cloning and physical maps of plasmid DNAs from *Streptomyces noursei, J. Antibiot.,* 36, 1743, 1983.
352. **Thompson, C. J., Kieser, T., Ward, J. M., and Hopwood, D. A.,** Physical analysis of antibiotic-resistance genes from *Streptomyces* and their use in vector construction, *Gene,* 20, 51, 1982.
353. **Thompson, C. J., Ward, J. M., and Hopwood, D. A.,** DNA cloning in *Streptomyces:* resistance genes from antibiotic-producing species, *Nature (London),* 286, 525, 1980.
354. **Thompson, C. J., Ward, J. M., and Hopwood, D. A.,** Cloning of antibiotic and nutritional genes in *Streptomyces, J. Bacteriol.,* 151, 668, 1982.
355. **Townsend, M. E., Wright, H. M., and Hopwood, D. A.,** Efficient mutagenesis by near ultraviolet light in presence of 8-methoxypsoralen in *Streptomyces, J. Appl. Bacteriol.,* 34, 799, 1971.
356. **Toyama, H., Okanishi, M., and Umezawa, H.,** Physical characteristics of plasmids from *Streptomyces kasugaensis* MB273, *Plasmid,* 5, 306, 1981.
357. **Toyama, H., Hayashi, E., Nojiri, C., Katsumata, K., Miyata, A., and Yamada, Y.,** Isolation and characterization of small plasmids from *Streptomyces, J. Antibiot.,* 35, 369, 1982.
358. **Toyama, H., Hayashi, E., Hoshiko, S., Nagaoka, K., and Yamada, Y.,** Isolation and characterization of new virulent actinophage ϕr5, *Agric. Biol. Chem.,* 47, 1853, 1983.
359. **Toyama, H., Hayashi, E., Nagaoka, K., and Yamada, Y.,** Protoplast transfection of *Streptomyces chartreusis* SF1623 with actinophage ϕr5 DNA, *Agric. Biol. Chem.,* 47, 1859, 1983.
360. **Troost, T. R., Danilenko, V. N., and Lomovskaya, N. D.,** Fertility properties and regulation of antimicrobial substance production by plasmid SCP2 of *Streptomyces coelicolor, J. Bacteriol.,* 140, 359, 1979.
361. **Umezawa, H., Okami, Y., and Hotta, K.,** Transfer of leupeptin-producing ability of the strain, *Streptomyces roseus* MA839-A1, by conjugation, *J. Antibiot.,* 31, 95, 1978.
362. **Unowsky, J. and Hoppe, D. C.,** Increased production of the antibiotic aurodox (X-5108) by aurodox-resistant mutants, *J. Antibiot.,* 31, 662, 1978.
363. **Upshall, A., Giddings, B., Teow, S. C., and Mortimer, I. D.,** Novel methods of genetic analysis in fungi, in *Third International Symposium on the Genetics of Industrial Microorganisms,* Sebek, O. K. and Laskin, A. I., Eds., American Society for Microbiology, Washington, D.C., 1979, 197.
364. **Vandamme, E. J., Leyman, D., DeBuyser, D. R., De Visscher, P. L., Spriet, J. A., Vansteenkiste, G. C., Nimi, O., Poirier, A., and Demain, A. L.,** Environmental influences on the dynamics of the gramicidin S fermentation, in *Peptide Antibiotics — Biosynthesis and Functions,* Kleinkauf, H. and von Döhren, H., Eds., de Gruyter, Berlin, 1982, 117.
365. **Vining, L. C.,** Antibiotic tolerance in producer organisms, *Adv. Appl. Microbiol.,* 25, 147, 1979.
366. **Vining, L. C.,** Peptides, in *Annual Reports on Fermentation Processes,* Vol. 4, Tsao, G. T., Ed., Academic Press, New York, 1980, 123.
367. **Vining, L. C. and Westlake, D. W. S.,** Chloramphenicol: properties, biosynthesis and fermentation, in *Biotechnology of Industrial Antibiotics,* Vandamme, E. J., Ed., Marcel Dekker, New York, 1984, 387.
368. **Vining, L. C. and Wright, J. L. C.,** Biosynthesis of oligopeptides, in *Biosynthesis,* Vol. 5, Bu'Lock, J. D., Ed., Chemical Society, London, 1977, 240.
369. **Vivian, A.,** Genetic control of fertility in *Streptomyces coelicolor* A3(2): plasmid involvement in the interconversion of UF and IF strains, *J. Gen. Microbiol.,* 69, 353, 1971.
370. **Vivian, A. and Hopwood, D. A.,** Genetic control of fertility in *Streptomyces coelicolor* A3(2): the IF fertility type, *J. Gen. Microbiol.,* 64, 101, 1970.
371. **Vivian, A. and Hopwood, D. A.,** Genetic control of fertility in *Streptomyces coelicolor* A3(2): new kinds of donor strains, *J. Gen. Microbiol.,* 76, 147, 1973.
372. **Voyeykova, T. A., Orekhov, A. V., and Rebentish, B. A.,** New approaches to the study of restriction and modification systems in actinomycetes, *Actinomycetes,* 15, 152, 1980.
373. **Waksman, S. A.,** What is an antibiotic or an antibiotic substance?, *Mycologia,* 39, 565, 1947.
374. **Waksman, S. A., Reilly, H. C., and Johnstone, D. B.,** Isolation of streptomycin-producing strains of *Streptomyces griseus, J. Bacteriol.,* 52, 393, 1946.
375. **Wang, Y.-G., Davies, J. E., and Hutchinson, C. R.,** Plasmid DNA in the erythromycin-producing microorganism, *Streptomyces erythreus* NRRL-2338, *J. Antibiot.,* 35, 335, 1982.
376. **Weber, J. M., Wierman, C. K., Hutchinson, C. R., Wang, Y.-G., and Davies, J.,** Genetics and biochemistry of erythromycin production in *Streptomyces erythreus* NRRL2338, in *Trends in Antibiotic Research. Genetics, Biosyntheses, Actions and New Substances,* Umezawa, H., Demain, A. L., Hata, T., and Hutchinson, C. R., Eds., Japan Antibiotics Research Association, Tokyo, 1982, 55.
377. **Weisblum, B., Fjuisawa, Y., and Horinouchi, S.,** Inducible MLS resistance in *Streptomyces* which synthesize MLS antibiotics: relation to antibiotic production and enhanced antibiotic yield, in *Trends in Antibiotic Research: Genetics, Biosyntheses, Actions and New Substances,* Umezawa, H., Demain, A. L., Hata, T., and Hutchinson, C. R., Eds., Japan Antibiotics Research Association, Tokyo, 1982, 73.

378. **Welsch, M.,** Lysogenicity in streptomycetes, *Ann. N.Y. Acad. Sci.*, 81, 974, 1959.
379. **Westley, J.,** Polyether antibiotics: versatile carboxylic acid ionophores produced by *Streptomyces, Adv. Appl. Microbiol.*, 22, 177, 1977.
380. **Westley, J.,** Polyether antibiotics — biosynthesis, in *Antibiotics*, Vol. 4, *Biosynthesis*, Corcoran, J. W., Ed., Springer, Berlin, 1981, 41.
381. **White, R. J.,** Anthracyclines, in *Biochemistry and Genetic Regulation of Commercially Important Antibiotics*, Vining, L. C., Ed., Addison-Wesley, Reading, Mass., 1983, 277.
382. **Whitney, J. G., Brannon, D. R., Mabe, J. A., and Wickner, K. J.,** Incorporation of labeled precursors into A16886B, a novel β-lactam antibiotic produced by *Streptomyces clavuligerus, Antimicrob. Agents Chemother.*, 1, 247, 1972.
383. **Williamson, S. E.,** The Isolation, Purification and Characterization of Four Restriction Endonucleases from Actinomycetes, M.Sc. thesis, Dalhousie University, 1980, 147 pp.
384. **Wright, L. F. and Hopwood, D. A.,** Identification of the antibiotic determined by the SCP1 plasmid of *Streptomyces coelicolor* A3(2), *J. Gen. Microbiol.*, 95, 96, 1976.
385. **Wright, L. F. and Hopwood, D. A.,** Actinorhodin is a chromosomally determined antibiotic in *Streptomyces coelicolor* A3(2), *J. Gen. Microbiol.*, 96, 289, 1976.
386. **Xue, Y., Dong, K., Li, M., and Zhu, Y.,** Genetic evidence of the presence of plasmid in *Streptomyces griseus* and its relationship with the biosynthesis of streptomycin, *Acta Microbiol. Sinica*, 18, 195, 1978.
387. **Yagisawa, M., Huang, T.-S. R., and Davies, J. E.,** Possible involvement of plasmids in biosynthesis of neomycin, *J. Antibiot.*, 31, 809, 1978.
388. **Yamashita, F., Hotta, K., Kurasawa, S., Okami, Y., and Umezawa, H.,** Antibiotic formation by interspecific protoplast fusion in streptomycetes and emergence of drug resistance by protoplast regeneration, *Abstr. 4th Int. Symp. Genet. Indust. Microorg.*, Kyoto, 1982, 108.
389. **Young, F. E. and Wilson, G. A.,** Genetics of *Bacillus subtilis* and other Gram-positive sporulating bacilli, in *Spores*, Vol. 5, Halvorson, H. O., Hanson, R., and Campbell, L. L., Eds., American Society for Microbiology, Washington, D. C., 1972, 77.
390. **Stuttard, C.,** Personal communication.
391. **Stuttard, C., Wong, K., and Doull, J.,** Personal communication.
392. **Schrempf, H. I. and Doull, J. L.,** Personal communication.

Chapter 4

FORMATION PHYSIOLOGY OF ERGOT ALKALOIDS

Zdeněk Řeháček

TABLE OF CONTENTS

I. INTRODUCTION

The divergence of individual directions in research has made it impossible for a single individual to comprehend the whole of a given field. Thus, it is important in each area to understand the stage of analytical development where it is possible and useful to begin at least partial synthesis. This is particularly true for ergot alkaloids, whose application and production depends on the cooperation of plant physiologists, microbiologists, biochemists, geneticists, organic chemists, pharmacologists, and biotechnologists.

As well as giving a brief review of recent literature, this paper also illustrates the present problems of ergot studies, with emphasis on areas where knowledge is lacking and where scientific endeavor should be pursued further. It is hoped that this contribution will also stimulate integration of individual fields of alkaloid research.

II. ERGOT ALKALOIDS

Ergot *(Secale cornutum)* is a remarkable compound with a fascinating history extending back over many centuries. It is the sclerotial or dormant overwintering form of *Claviceps* species which parasitizes the seed heads of grasses and cereals. Ergot was recognized as a fungus in 1711, but the complete life cycle was not described until 1853 by Tulasne. The ergot or ergot body is a hard, dark purple, sickle-shaped body in the seed head of various grains. The ergot body (sclerotium) falls to the ground and germinates the next spring (Figure 1). Short stalks terminated by a head (hence *Claviceps*) which contains the elongated ascospores grow from sclerotium. The ascospores are produced in a sexual process which involves the fusion of two nuclei and the subsequent formation of eight ascospores. The ascospores are released into the air, land on the moist stigmate of the flower, and germinate. The hyphae (sphacelia) grow around and then within the ovary of the flower. Asexual spores, or conidia, are produced and are spread to other flowers by honeydew produced by the host plant. In the final stage the hyphae become thicker and more closely packed to produce the hard sclerotium.

Sclerotium can be considered as the early stage of the sexual differentiation of *Claviceps*. Lacking conditions of the next sequence (cold treatment, water, respiration, or light), morphological specialization does not proceed, the sclerotium elongates, and enhanced activity will be demonstrated by a marked chemical specialization, i.e., intensive alkaloid synthesis.

Over the centuries, the role of *Claviceps* has undergone important changes from a dreaded toxic parasite to an important source of biologically effective substances. Ergotism, certainly the first described form of a mycotoxicosis, was recorded very early in history. Ergotism is the disease which results from consumption of the ergot body of sclerotium from rye or other grains infected by a parasitic fungus of the genus *Claviceps*. The disease is due to the effects of the alkaloids produced by the fungus and contained in the ergot body.

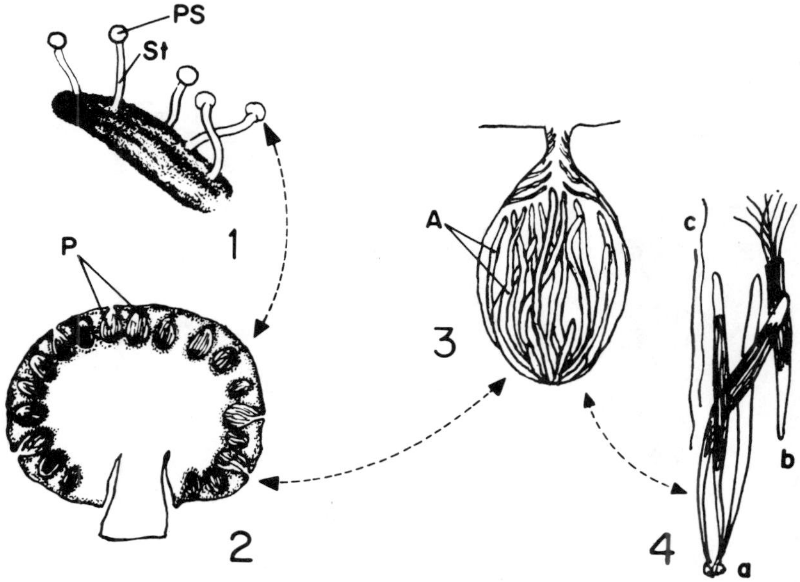

FIGURE 1. Germinating sclerotium (1) of *Claviceps purpurea;* 2 = stroma; P = perithecium; 3 = perithecium; A = asci; 4a = asci with ascospores; 4b = opened ascus with ascospores; 4c = ascospores.

At the beginning of the 1950s, the biosynthesis of ergot alkaloids drew a great deal of attention. It rapidly changed from a subject of fruitful speculation to one of experimetnal research. Primarily, the chemical properties of the alkaloids were studied. In the meantime, biologists were studying the physiological aspects of the synthesis of alkaloids from a practical point of view, i.e., for both agronomists, who were trying to obtain grain with the lowest possible alkaloid content, and for pharmacognosists, who were interested in high alkaloid yields. The dependence of the pharmacological effectiveness on the alkaloid chemical structure, enabling preparation of substances for specific purposes, was an important stimulus for the development of ergot alkaloid research.

The ergot alkaloids and their derivatives display such a diversity of biological activities that they cannot be regarded as a single pharmacological or therapeutic entity. There is no evidence to suggest that the diversity of biological properties exhibited by ergot compounds is attributable to a common underlying mechanism at the cellular or molecular level. It is more likely that their wide range of biological activities may best be explained by assuming:[1] (1) that ergot compounds interact with more than one receptor site; (2) that the population of receptor sites to which they have access varies from organ to organ; (3) that affinity for receptors sites and intrinsic activity vary from compound to compound and are determined by their chemical configuration.

The ergot alkaloids find application in the treatment of a variety of clinical conditions. Probably the most important of their many indications are postpartum hemorrhage, migraine and other vascular headaches, venostasis, orthostatic hypertension, senile cerebral insufficiences, conditions associated with hyperprolactinemia, and Parkinson's disease. Stimulation of uterine motor activity is the oldest therapeutic use of ergot alkaloids. Ergot alkaloids and their derivatives used in therapy are given in Figure 2.

Several semisynthetic compounds of the ergot class, including bromokryptin,[2] lergotrile,[3] pergoline, and 6-ethyl-9-oxaergoline (Figure 3) have proven to be dopamine agonists. The neuropharmacology of dopamine agonists has long been of interest due to their utility in treating diseases such as Parkinson's, galactorrhea, and hyperprolactinemia. The newer ergot derivatives have also been implicated as potential therapeutic agents in diseases states such

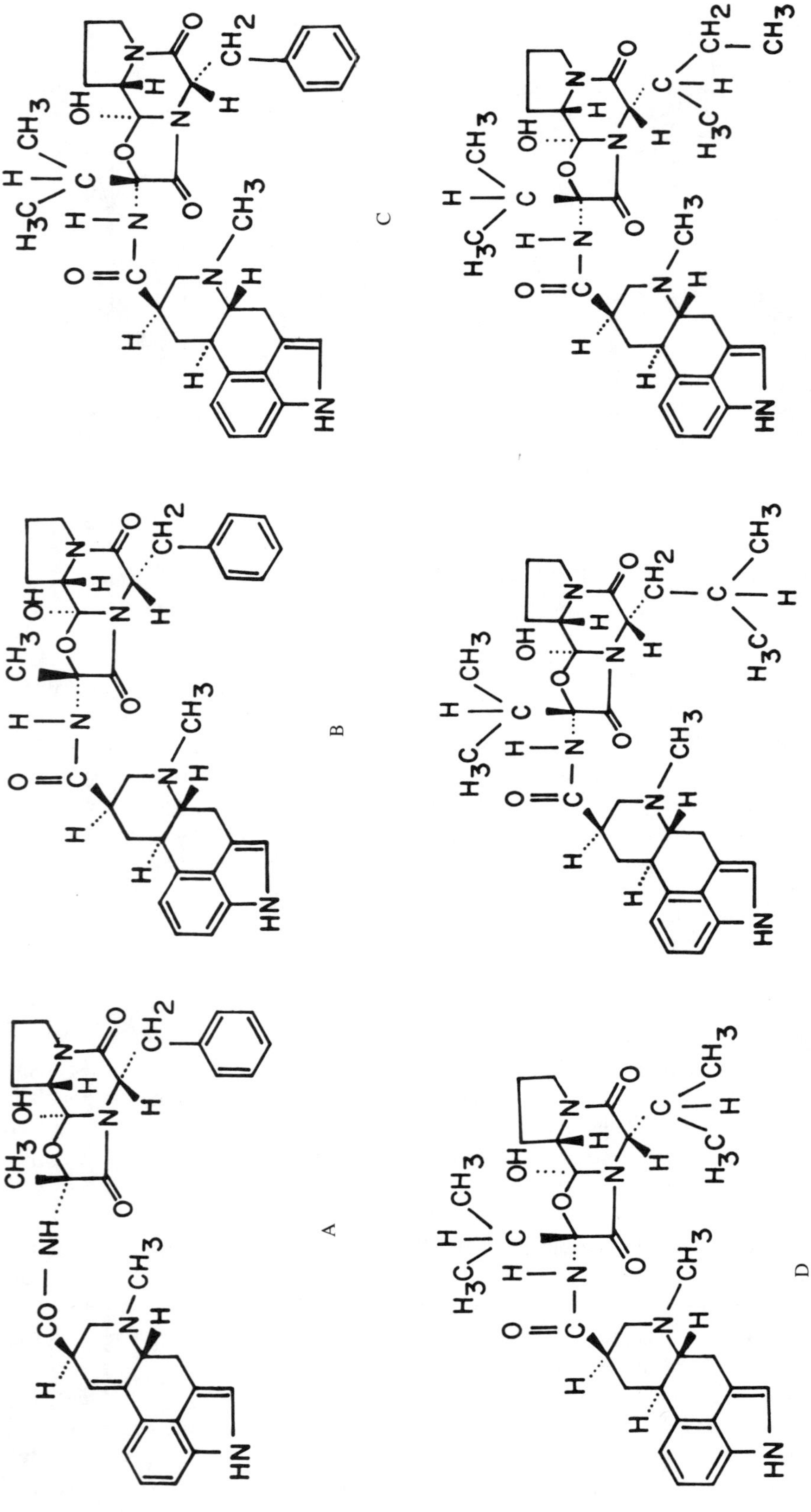

FIGURE 2. Some ergot compounds used in therapy. (A) Ergotamine; (B) dihydroergotamine; (C) dihydroergocristine; (D) dihydroergocornine; (E) dihydro-α-ergokryptine; (F) dihydro-β-ergokryptine; (G) ergobasine; (H) methylergobasine; (I) LSD-25; (J) methylsergide; (K) methergoline; (L) nicergoline; (M) bromokryptine; (N) lisuride.

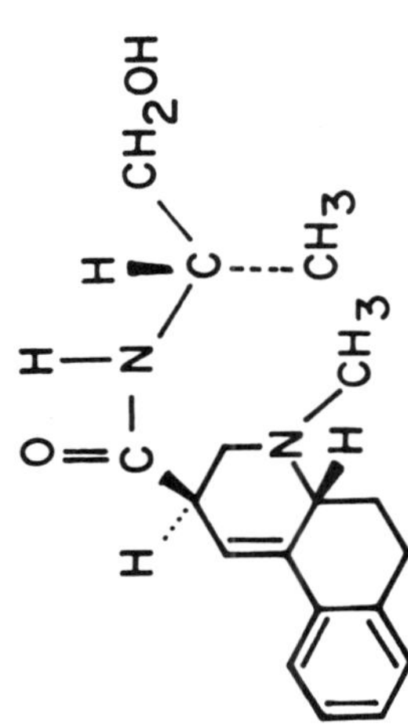

G

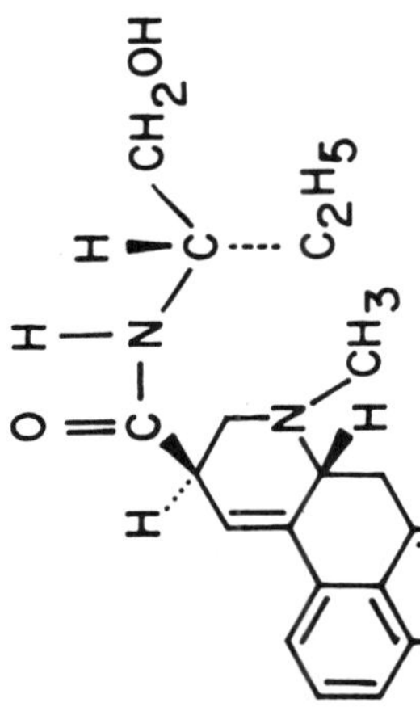

H

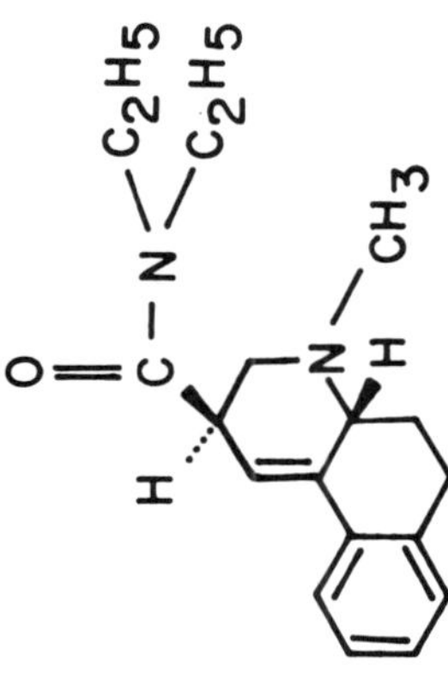

I

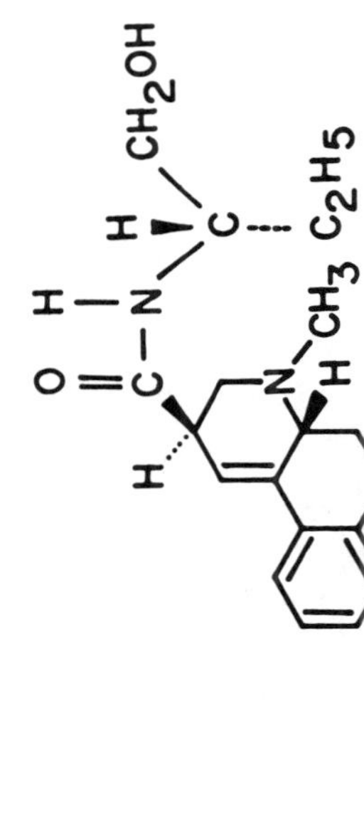

J

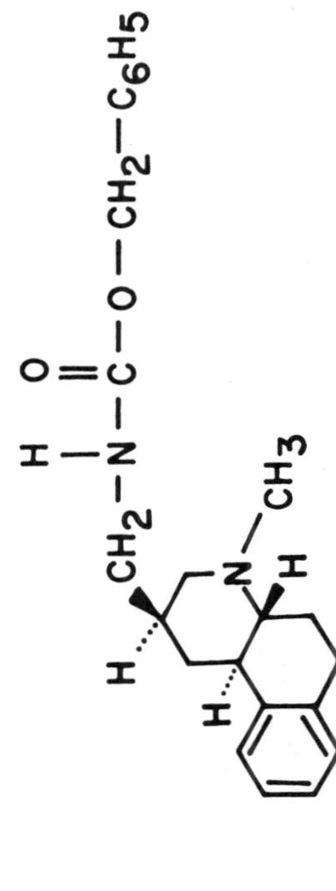

K

FIGURE 2 (continued)

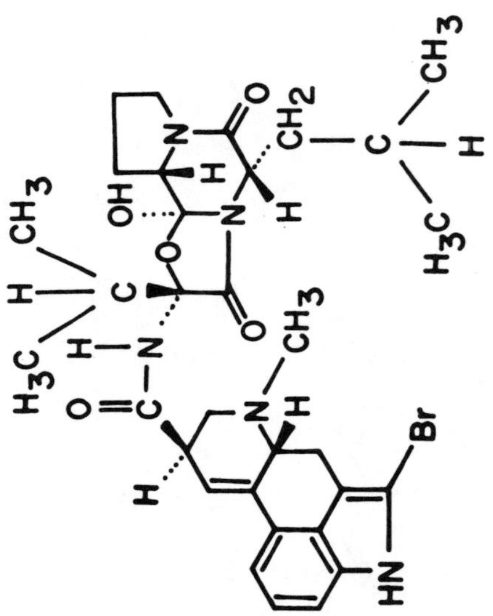

M

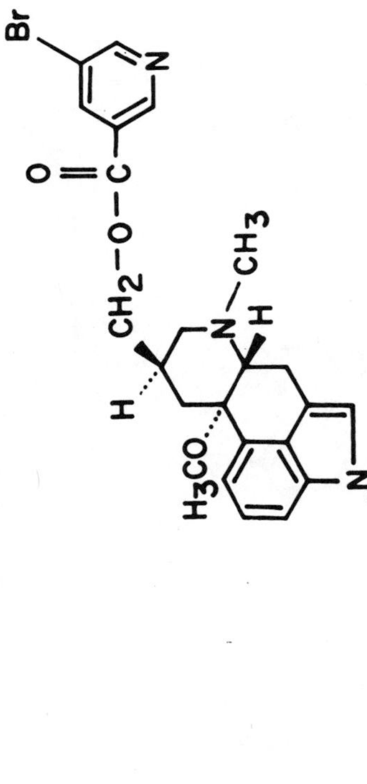

L

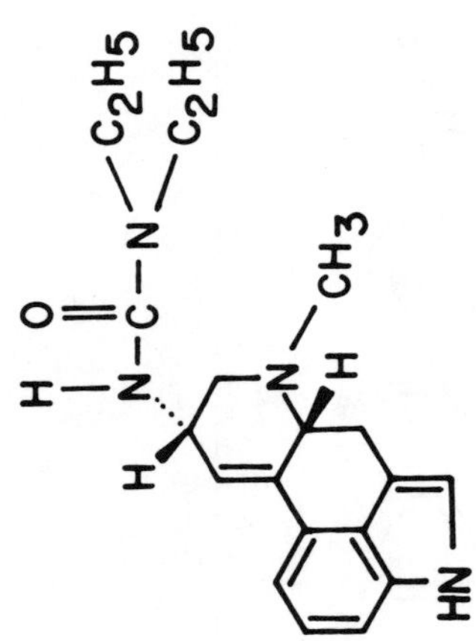

N

FIGURE 2 (continued)

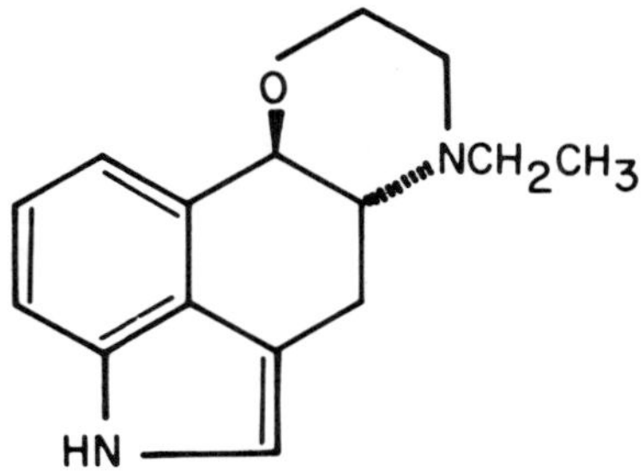

FIGURE 3. 6-Ethyl-9-oxaergoline, a novel potent dopamine agonist.[4]

as acromegaly, suppression of postpartum lactation, treatment of breast cancer, and possibly, cancer of the prostate (secondary to inhibition of prolactin secretion).

A. Living Sources

In nature, ergot alkaloids are formed primarily by the fungus *Claviceps*. More than 40 different alkaloids have been isolated from this pyrenomycete, which has about 50 species. An increasingly advanced development within the genus *Claviceps* has been demonstrated which reaches a maximum in *Claviceps purpurea*. It is interesting to note that the complexity of the formed alkaloids follows the evolutionary status of this species. *C. purpurea* readily infects rye, but also barley, wheat, and more than 100 grasses. Other species infect rice, wheat, millet, and corn.

The nonalkaloid indole derivatives identified in saprophytic cultures of *Claviceps* sp. are γ, γ-dimethylallyl-tryptophan,[5] clavicipitic acid, *N*-acetyltryptophan, 5-hydroxytryptophan, and a soluble brown pigment derived from tryptophan. Ergot alkaloids have also been found outside the genus *Claviceps* (Table 1) especially in *Aspergillus* and *Penicillium*.[6,7] Vining et al.[8] isolated from *Penicillium citreoviride* a new type of clavine alkaloid containing two ergoline ring systems. The dimer was called "cividiclavine" (Figure 4). It contains a pyroclavine moiety linked through its indolic nitrogen to a hydroxypyroclavine. The linkage in the latter is tentatively placed at C-13′ and the hydroxy group is assigned to C-14′. Free pyroclavine was also isolated from the culture. Several investigators showed a wider distribution of these alkaloids in the Convolvulaceae (Morning-Glory family). These currently include the genera *Rivea, Ipomoea, Argyreia, Cuscula,* and *Stictocardia*. The seeds of *Argyreia nervosa* contain the highest percentage of indole alkaloid constituents (0.5 to 0.9%) of the genera of the Convolvulaceae thus far studied. Ergine (0.13%) and isoergine (0.18%) were found in the highest concentration.

B. Production

Production of ergot alkaloids for drug use has progressed from the collecting of sclerotia formed by natural infection to systematic cultivation of *C. purpurea* in rye fields. However, field production is labor intensive and weather dependent, so that there has always been a strong interest in artificial culture of the ergot fungus. Attempts to produce the alkaloids in vitro met with little success until Abe[9] showed that good yields could be obtained with strains isolated from certain wild grass ergots. The cultures produced a new type of alkaloid and did not immediately lead to commercial fermentations.

Sclerotia of *C. purpurea,* i.e., erogt *(Secale cornutum),* collected from rye for the extraction of peptide alkaloids may contain 0.8% product by dry weight. This content would be the equivalent of about 500 mg/ℓ alkaloid level in a fermentation broth of 6% mycelial dry weight. The production of ergot alkaloids is today largely carried through submerged culture of strains of the genus *Claviceps*. The species most commonly used for this purpose

Table 1
ALKALOIDS OUTSIDE THE GENUS *CLAVICEPS*

Organism	Alkaloid
Sphacelia sorghi	Dihydroergosine
Penicillium concavorugulosum	Chanoclavine-I
	Rugulovasine A and B
P. roquefortii	Festuclavine
	Roquefortine A, B, C
	Isofumigaclavine A
P. chermesinum	Costaclavine
P. fumigatus	Festuclavine
	Agroclavine
	Elymoclavine
	Chanoclavine
	Fumigaclavine A, B, C
Ipomoea hildebrandtii (seeds)	Cycloclavine
	Festuclavine
I. argyrophylla	Ergosine
Argyreia nervosa	Ergine
	Ergometrine
	Lysergic acid α-hydroxyethyl amide
	Clavines

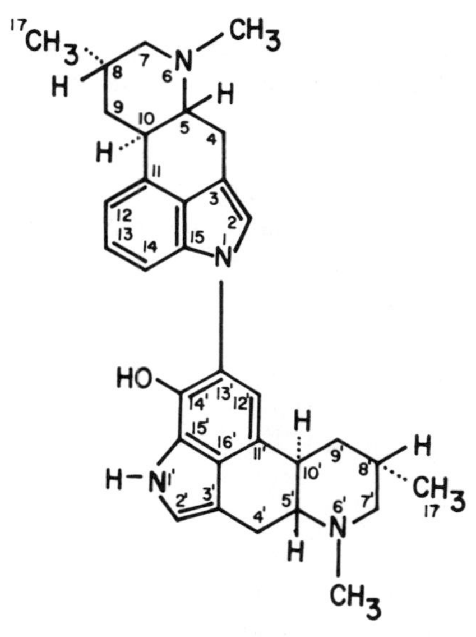

FIGURE 4. Cividiclavine.[8]

are listed in Table 2. Published data of peptide alkaloid fermentations carried out with laboriously developed industrial strains refer to levels not higher than 1000 to 1600 mg/ℓ; the simultaneous increase in mycelial yield allows a maximum of about 2% alkaloid of the cell dry weight.

A number of syntheses of lysergic acid have been recorded. All have avoided the use of the biosynthetic precursor, tryptophan. Rebek and Tai[10] successfully converted this amino

Table 2
***CLAVICEPS* SPECIES MOST COMMONLY USED FOR SUBMERGED FERMENTATION OF ERGOT ALKALOIDS**

Species	Alkaloid
C. fusiformis	Agroclavine
C. paspali	Lysergic acid α-hydroxyethylamide
	$\Delta^{8,9}$ Lysergic acid
	Ergometrine
C. purpurea	Elymoclavine
	Ergosine
	Ergotamine
	α-Ergokryptine
	β-Ergokryptine
	Ergocornine
	Ergocristine

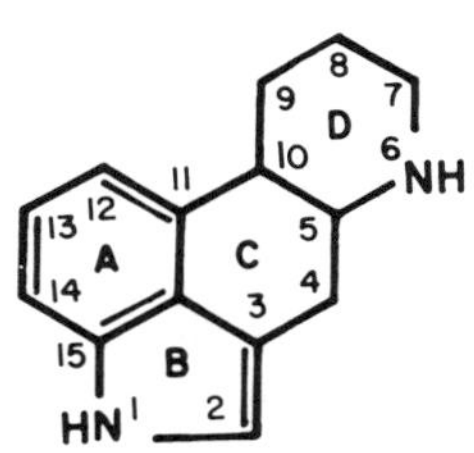

FIGURE 5. Ergoline.

acid to the ergot alkaloids rugulovasine and lysergine, and elaborated a new efficient synthesis of lysergic acid in ten steps.

Partial alkaloid synthesis is employed in studying the structure and pharmacological effect of peptide alkaloids with various peptide parts of the molecule. The total synthesis of lysergic acid, the solution of the spatial arrangement of this substance, the synthesis of the peptide part of some alkaloids, and especially the total synthesis of ergotamine, allow the synthesis of most ergot alkaloids with very important results.[11] However, total synthesis remains uneconomical.

The ergot alkaloids can be chemically manipulated to form derivatives that possess altered pharmacological effects. The selective saturation of one of the double bonds (C9 to C10) of the lysergic acid moiety results in compounds which differ considerably from the natural alkaloids. One of the most prominent divergencies was found in respect to the uterine effect. The dihydrogenated alkaloids lost the excitatory effect on the uterus, but they are indeed able to inhibit in vitro and *in situ* the powerful stimulative action of the natural alkaloids such as ergotamine and ergotoxine. In the preparation of practically important nicergoline,[12] attention was paid to the modification of the ergoline rest.

C. Chemistry

The erogt alkaloids constitute the largest known group of nitrogenous fungal metabolites. Their chemistry has been reviewed in several works.[13-15] A common part of ergot alkaloids is ergoline, i.e., partially hydrogenated indolo (4,3-fg)quinolone (Figure 5). Most natural ergot alkaloids form 6,8-dimethyl-$\Delta^{8,9}$ or $\Delta^{9,10}$-ergolene derivatives. Ergot alkaloids can be separated into two basic groups: derivatives of lysergic acid and clavine alkaloids. Bu'Lock and Barr[16] also include tricyclic chanoclavines in the clavine group.

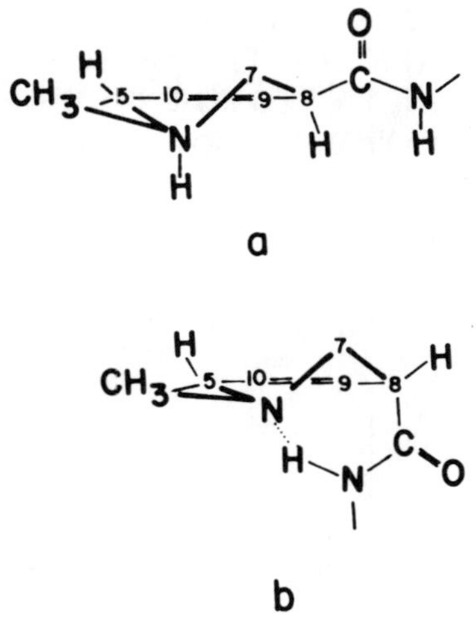

FIGURE 6. The D-ring partial conformation of derivatives of lysergic (a) and isolysergic (b) acids.

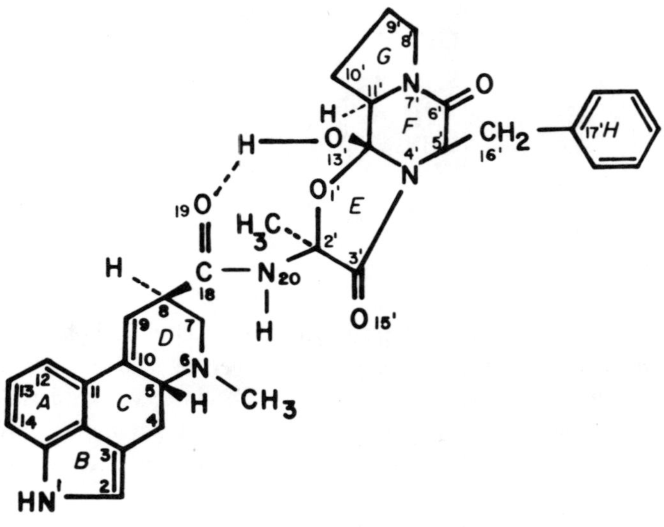

FIGURE 7. Ergopeptine representative (ergotamine).

The derivatives of lysergic acid so far isolated are the amides, in which the amide part is a small peptide or a simple alkylamide. The basicness of the amides of lysergic acid is a result of the presence of nitrogen in the 6-position. The pharmacologically active levorotatory lysergic acid series and the inactive dextrorotatory lysergic acid series differ only in configuration at C-8 (Figure 6); both have trivial names beginning with ''ergo''. The epimers are distinguished by the suffixes ''ine'' and ''inine'', respectively. The clavine series usually has trivial names with ''clavine'' as a suffix. All naturally occurring alkaloids except one enantiomer of chanoclavine-II have the R configuration at C-5 and are designated as d-forms.

The classic peptide ergot alkaloids (ergopeptines) (Figure 7) are characterized by a modified tripeptide containing L-proline[17] and α-hydroxy-α-amino acid, which formed cyclol with a carbon from the proline carboxylic group.[18,19] Erogpeptines are the only natural substances with demonstrated cyclol structure. Alkaloids with hydrogen in position 5 and

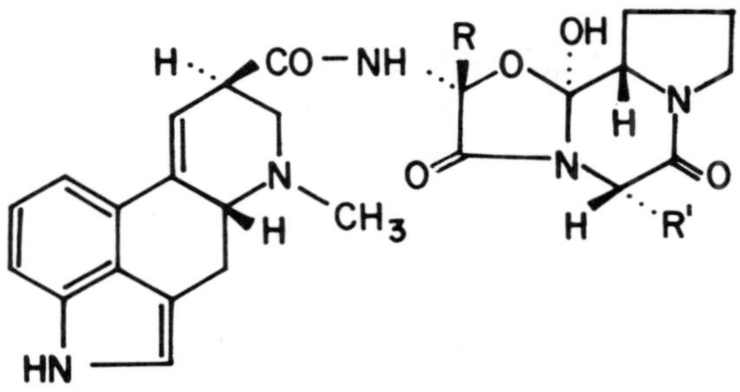

FIGURE 8. Ergopeptine.

Table 3
GROUPS OF GENUINE ERGOPEPTINES[20]

R′	Ergotamine $R{=}CH_3$	Ergoxine $R{=}C_2H_5$	Ergotoxine $R{=}CH(CH_3)_2$
$CH_2{-}C_6H_5$	Ergotamine	Ergostine	Ergocristine
	Ergotaminine	Ergostinine	Ergocristinine
$CH_2{-}CH(CH_3)_2$	α–Ergosine	α-Ergoptine	α-Ergokryptine
	α-Ergosinine	α-Ergoptinine	α-Ergokryptinine
$CHCH_3{-}C_2H_5$	β-Ergosine	β-Ergoptine	β-Ergokryptine
	β-Ergosinine	β-Ergoptinine	β-Ergokryptinine
$CH(CH_3)_2$	Ergovaline	Ergonine	Ergocornine
	Ergovalinine	Ergoninine	Ergocorninine
$CH_2{-}CH_3$	(Ergobine)[1]	Ergobutine[1]	Ergobutyrine[1]

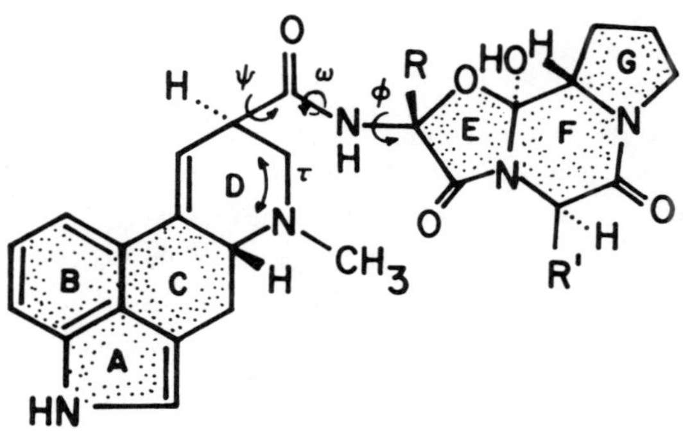

FIGURE 9. Rigid (shaded ring systems) and flexible parts of ergopeptine.[21]

10 which are spatially alternately oriented are the subject of considerable pharmacological interest.

Natural ergopeptines have been divided into three groups, each of them having the same substituent (R) at position 2′ of the peptide moiety (Figure 8 and Table 3). These alkaloids may also be divided into series having the same substituent (R′) at position 5′. The series at present only lacks the alkaloid of the ergotamine group, for which, however, Bianchi et al.[20] propose the name ''ergobine''. Studies into the isolation of this alkaloid are in progress.

It can be easily seen by model building (Figure 9) that the ergopeptine molecule consists of two rigid ring systems linked together by two flexible fragments: the D ring and the central amide. Free rotation is theoretically possible around two single bonds (ϕ, ψ), restricted rotation around the partial double bond CO − NH (ω), and pseudo-rotation in the D ring

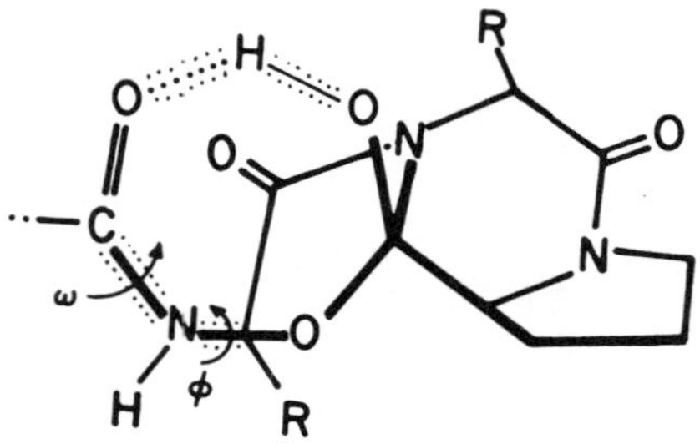

FIGURE 10. Conformation of the central amide linked to the peptide backbone by the intramolecular hydrogen bond from the cyclol-hydroxyl to the amide carbonyl.[21]

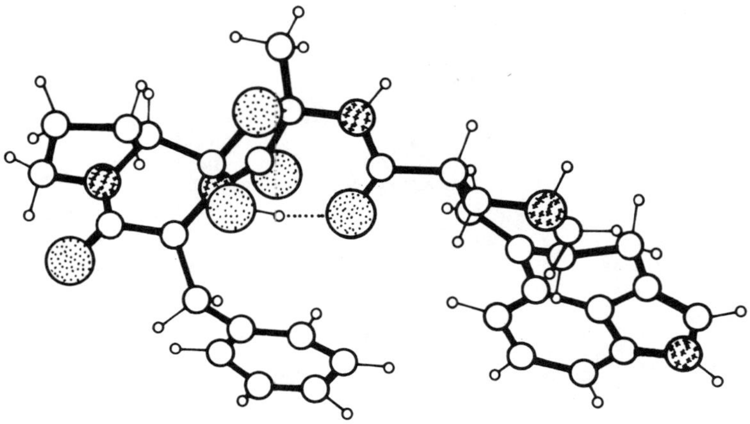

FIGURE 11. The crystalline molecular conformations of ergotamine.[21]

(τ).[21] As indicated, the overall molecular conformation can essentially be described with only four angles: the torsion angle τ, defining the conformation of ring D; two torsion angles, ϕ and ψ, for the rotation around the two single bonds to the central amide; the torsion angle ω (= 0 or 180°), defining the *cis/trans*-configuration of the partial double bond CO – NH. Taking into account some steric constraints in the rotations around the bonds, i.e., H....H repulsions, the number of theoretically possible conformations appears to be limited, but still quite large nevertheless.

Figure 10 shows a schematic drawing of the conformation of the central amide linked to the peptide backbone by the intramolecular hydrogen bond from the cyclol-hydroxyl to the amide carbonyl. This partial conformation, which fixes both the *trans*-configuration of the amide and the torsion angle ϕ, has been observed in all ergopeptines investigated so far.

The crystalline molecular conformation of the ergopeptine representative (ergotamine) is shown in Figure 11.

D. Separation and Measurement

Chemical assays of ergot alkaloids are usually based on reaction with *p*-dimethylaminobenzaldehyde in which most ergolines give an intense blue color and show similar molar absorbances. To separate natural alkaloid mixtures into their components, a variety of chromatographic methods have been elaborated.

A method of high performance liquid chromatography (HPLC) has been developed for the separation and quantitative analysis of a mixture of ergot alkaloids.[22,23] Four isomers of lysergic acid α-hydroxyethylamide and its decomposition products, ergine and erginine, can be separated by the HPLC method and identified by spectroscopy.[24] This HPLC method can be used to study kinetics of isomerization reactions and degradation reactions of lysergic acid α-hydroxyethylamide as well as for the control of the purity of preparations.[25]

Mass spectra of ergopeptines characterize the peptide part of the molecule. Its ionic fragments represent 80 to 90% of the total ion current. Ion *c* corresponding to the complete peptide group decomposes in an ionic sequence which is influenced by substitution in the $C_{(5')}$ position. The molecular peaks, when present, and some characteristic fragment peaks, facilitate a quick identification of ergot alkaloids by means of mass spectrometry.[26]

E. Medicinal Chemistry

Medicinal chemistry has evolved from simply "organic drug synthesis" to a multidisciplinary approach toward the rational synthesis of therapeutic agents, based on an understanding of the interaction of the drug with the biological system. Structure-activity relationships represent the elucidation of how structural modification of the parent drug alters its biological action (potency, toxicity). Since the metabolism of many drugs yields products that are also active, structure-metabolic relationships are becoming increasingly more important. Elucidation of the mechanism of action clearly aids in the design of newer and better agents. A final area and one of paramount importance, is metabolism. Certain drugs are inactive as administered and must undergo bioactivation before the therapeutic effect can be realized. Conversely, drug action must be terminated, and bioinactivation is usually responsible. Further, it is well recognized that products may result that exhibit toxicity to specific organs. The more the biochemist knows about predicting routes of metabolism, the better his position to control desired routes and avoid undesirable routes of metabolism.

The major medicinal interest in ergolines (lergotrile, pergolide) stems from their potent dopaminergic activity. The major route of lergotrile metabolism examined in the fungi and streptomycetes was *N*-dealkylation. The reaction went to completion with *Streptomyces platensis* NRRL 2364.[27] This parallels the major route of metabolism in mammals and also represents a mild, high-yielding alternative to chemical *N*-dealkylation.

With pergolide, *Helminthosporium* sp. strain NRRL 4761 yields only the sulfoxide and is capable of the reverse reaction, sulfoxide reduction to the sulfide (a type reaction also occurring in mammals). *Aspergillus alliaceus* V 1-315 represents a more complete mammalian model as it yields both the sulfoxide and the sulfone.[28] Both organisms exploit the idea of a predictable "type-reaction approach", since they were chosen on the basis of their reported abilities to carry out these sulfur oxidations.

III. BIOGENESIS

A. Ergoline

The precursors channeled into ergoline biosynthesis from primary metabolism are L-tryptophan, dimethylallylpyrophosphate, and a one-carbon doror, i.e., *S*-adenosyl-L-methionine (Figure 12). The ysergic acid amide derivatives require additional amino acids. Construction of ergoline from its biogenetic precursors begins with the isoprenylation of L-tryptophan at C-4. Chanoclavine-I is the first main sequence alkaloid to accumulate in cultures, and its formation from dimethylallyltryptophan requires decarboxylation, *N*-methylation, and hydroxylation. Isochanoclavine-I and the chanoclavine-II enantiomers are end products of irreversible branching reactions.

B. Clavines

The biogenesis of clavine alkaloids is illustrated in Figure 13. Chanoclavine-I is incorporated into elymoclavine with loss of the pro-17R and retention of the pro-17S hydrogen.

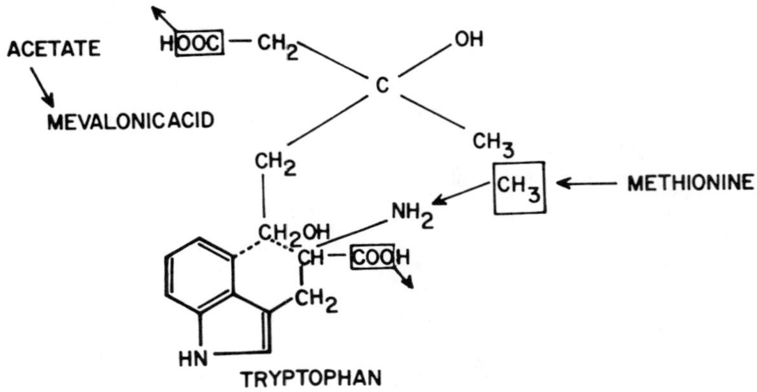

FIGURE 12. Biogenetic origins of the ergoline ring system in *Claviceps* sp.

The stereochemistry for hydrogen atom removal in the conversion of chanoclavine-I into elymoclavine is the same as for the yeast and liver alcohol dehydrogenases.[29] Chanoclavine-I aldehyde is an intermediate in the conversion of chanoclavine-I into elymoclavine. Setoclavine, isosetoclavine, penniclavine, and isopenniclavine are derived from agroclavine and elymoclavine. Both isomers, i.e., setoclavine-isosetoclavine and penniclavine-isopenniclavine, are formed in most fungal species that carry out the conversion. Thus, the respective hydroxylation seems to be nonenzymic.[30]

Clavine alkaloids lie at the beginning of the biogenetic order and classic peptide ergot alkaloids (ergopeptines) at the end. The oxidation order agroclavine-elymoclavine-lysergic acid is the main biogenetic pathway. The variety of clavine alkaloids is due to the action of enzymes that direct elymoclavine and agroclavine from the main biosynthetic route. Such shunt activity is particularly evident in strains that lack the complete pathway and are unable to synthesize substituted lysergic acids. Figure 14 shows the biogenesis of simple derivatives of lysergic acid.

C. Ergopeptines

Ergopeptines are the only natural substances with demonstrated cyclol structure. Kim and Anderson[31] proposed a paspalyl-enzyme intermediate in ergopeptine biosynthesis. The in vivo and in vitro incorporation experiments[32] indicate that lysergic acid is not an intermediate in the biosynthesis of the peptide ergot alkaloids, although added lysergic acid is incorporated. Some results indicate that paspalic acid is not an intermediate in peptide alkaloid biosynthesis. The experimental data of Kim and Anderson[33] are consistent with the sequence: elymoclavine is oxidized to elymoclavine aldehyde, elymoclavine aldehyde binds to enzyme, and oxidation occurs to give paspalyl-enzyme. In vivo the paspalyl group would be transferred (with migration of the $\Delta^{8,9}$ double bond to $\Delta^{9,10}$ to give lysergyl) to the amino group of, e.g., alanine in subsequent steps of peptide alkaloid biosynthesis.

Several biosynthetic routes to peptide antibiotics not utilizing mRNA as a primary template have been evaluated during the past years. Some routes have been shown to involve multienzyme types of structures. These have been described as multienzyme complexes, but many of them do not contain subunits, and are composed of multifunctional polypeptide chain carrying several catalytic functions.[34] Sequence and specificity of the amino acids in the peptide are determined by arrangement of thiols in the catalytic functions of the synthetases. This novel type of peptide biosynthesis is termed ''multienzyme thiotemplate mechanism''.[35] The synthetases are highly complicated enzyme systems and can be considered the most complex multienzymes known today.

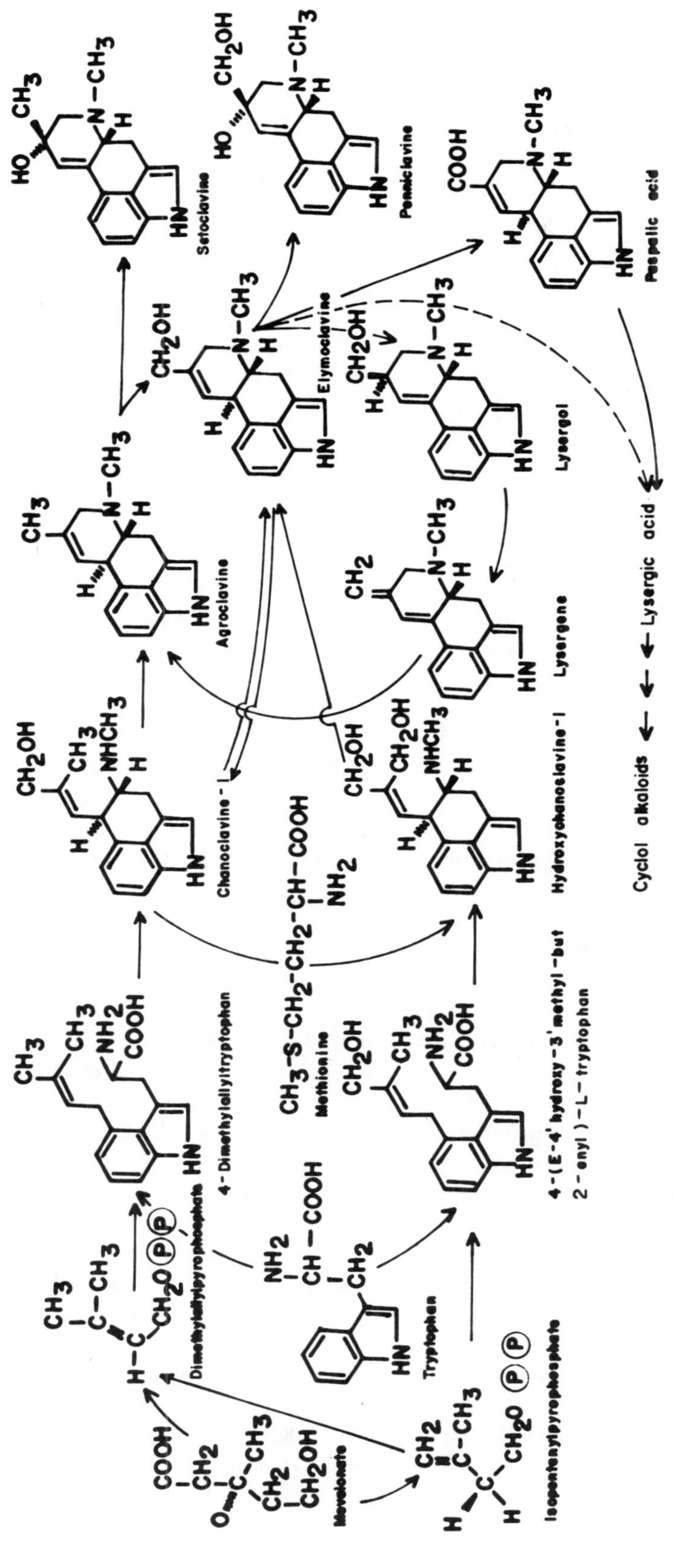

FIGURE 13. Biogenesis of clavine alkaloids.

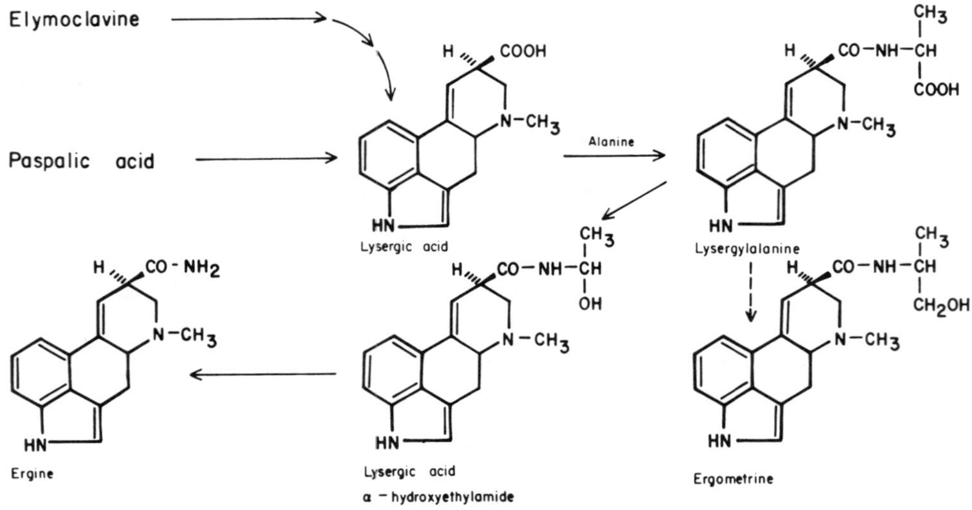

FIGURE 14. Biogenesis of simple derivatives of lysergic acid.

The formation of the peptide chain of ergoline begins with prolin (Figure 15).[36] No free intermediates occur in the biosynthetic pathway of ergopeptines. It is therefore conceivable that the synthesis of the peptide alkaloids takes place by mediation of a multienzyme complex.[37] The isolation and characterization of such an enzyme system in *Claviceps* has not yet been accomplished.

It seems that the activating enzyme and the enzymes that catalyze synthesis of the peptide moiety can tolerate some change in the lysergic acid structure. Dihydrolysergic acid added to the culture medium of an ergotamine-producing strain was converted to dihydroergotamine.[38] Exogenously administrated amino acids strongly influence the alkaloid spectrum in the maturing sclerotia under parasitic conditions.[39] In the case of ergotoxine alkaloids, the addition of any of the amino acids L-valine, L-leucine, and L-isoleucine to the nutrient medium significantly determines the proportions of the corresponding alkaloids ergocornine, α-ergokryptine, and β-ergokryptine. Analogs of peptide ergot alkaloids can be obtained by feeding a producing culture of *C. purpurea* with an analog of one of the amino acids of the peptide chain of the alkaloid[40] (Table 4). Baumert et al.[41] obtained an ergosine analog with a modified proline moiety by feeding an appropriate *Claviceps* strain with L-thiazolidine-4-carboxylic acid. This is the first directed biosynthesis of an ergot alkaloid with an altered proline moiety.

The amino acids present in the side chains of the 12 known ergot alkaloids comprise only 7 out of the 20 known natural amino acids. They are alanine, phenylalanine, leucine, isoleucine, valine, aminobutyric acid, and proline. One of the characteristics shared by these seven amino acids is the lack of polar groups in their side chains. It seems, therefore, that the biosynthetic enzymic system is able to discriminate between the amino acids on the basis of their side chain polarity.[42] Proline in a peptide is often chosen because it causes bends in the protein and is thought to be a marker of regions that exit closer to the surface of a protein molecule.

The biosynthesis of the peptide moiety of the ergot alkaloids is controlled by the relative concentration of amino acids in the internal pool, and point to the possibility of obtaining new and pharmacologically interesting alkaloids.[40,43] The competition seems to be restricted to structually related amino acids. e.g., branched chain amino acids or aromatic amino acids and their analogs, but not *inter alia.*

Peptides are not precursors of the peptide part of ergopeptines. Added peptides are split to the amino acids before incorporation into alkaloids. Condensation of the ergopeptine

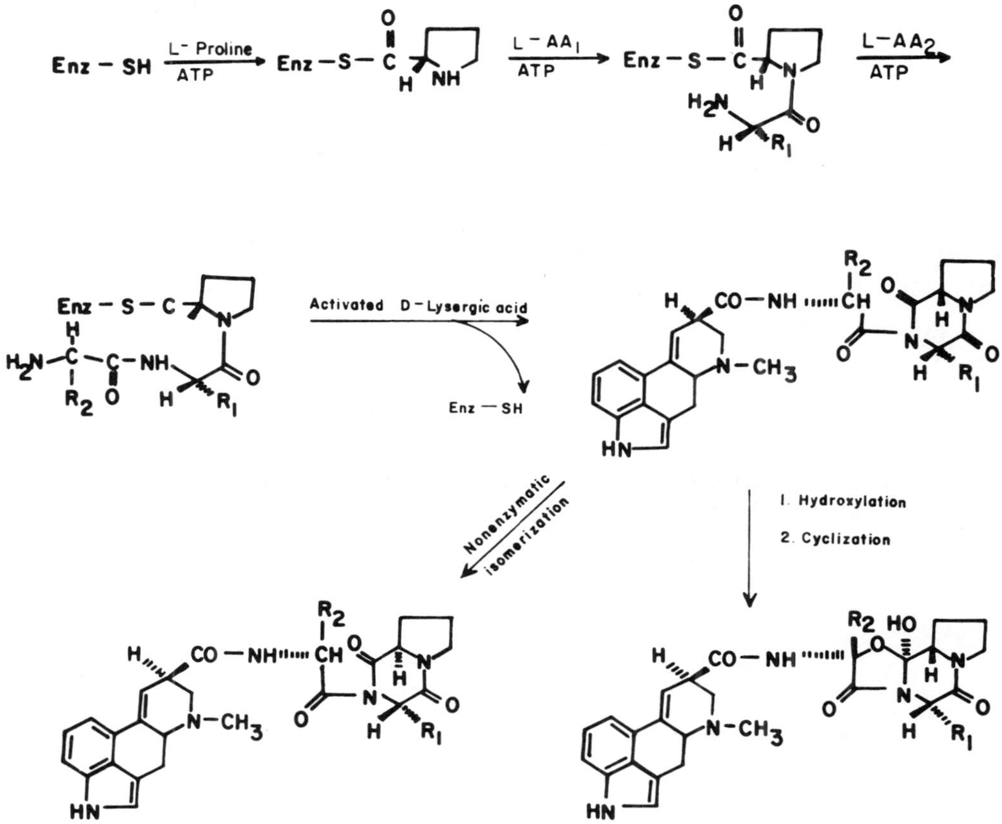

FIGURE 15. Formation of ergopeptines and ergopeptams.

Table 4
GROUPS OF GENUINE AND HYPOTHETIC ERGOPEPTAMS[49-51]

R′	Ergotamam $R{=}CH_3$	Ergoxam $R{=}C_2H_5$	Ergotoxam $R{=}CH(CH_3)_2$
$CH_2{-}C_6H_5$	Ergotamam	Ergostam	Ergocristam
	Ergotaminam	Ergostinam	Ergocristinam
$CH_2{-}CH(CH_3)_2$	α-Ergosam	α-Ergoptam	α-Ergokryptam
	α-Ergosinam	α-Ergoptinam	α-Ergokryptinam
$CHCH_3{-}C_2H_5$	β-Ergosam	β-Ergoptam	β-Ergokryptam
	β-Ergosinam	β-Ergoptinam	β-Ergokryptinam
$CH(CH_3)_2$	Ergovalam	Ergonam	Ergocormam
	Ergovalinam	Ergoninam	Ergocorninam

precursors occurs through the activated intermediates, e.g., CoA-thiolester of lysergic acid.[44] The "cyclol synthetase" seems to be not more than 48 hr stable in vivo.[45]

Rao and Singh[46] suggest that oligopeptides liberated during normal protein degradation may possess a part of the sequence/structure which, on undergoing reactions like cyclization, transpeptidation, and/or other modifications, can be converted into physiologically active peptides. In this situation also, a bioactive peptide is formed without the necessity of a specific precursor protein. The continual degradation of the numerous proteins in cells may

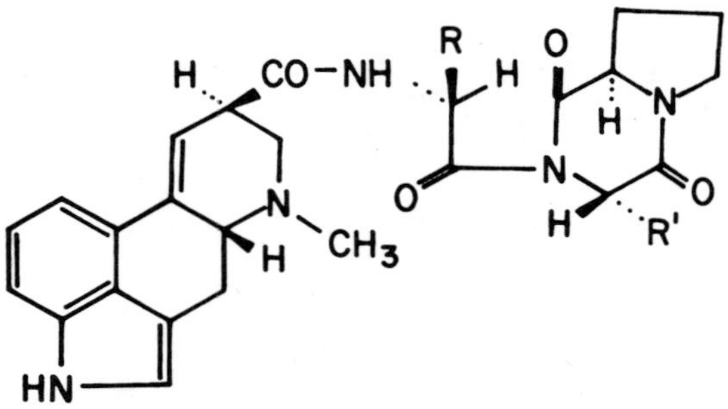

FIGURE 16. Ergopeptam.

Table 5
INCORPORATION OF AMINO ACID ANALOGS INTO ERGOPEPTINES BY VARIOUS *CLAVICEPS PURPUREA* STRAINS[40]

Strain	Main alkaloid produced	Amino acid or analog added to the medium	Amount of alkaloid with added amino acid or analog (% of the total)
S 40 phe-	Ergocristine	*p*-Chlorophenylalanine	63
		p-Fluorophenylalanine	66
S 40	Ergocristine	*p*-Chlorophenylalanine	35
		p-Fluorophenylalanine	35
275 FI	Ergotamine	*p*-Fluorophenylalanine	30
B 21	Ergocornine	L-Norvaline	59
	Ergokryptine	L-Norleucine	23
B 21 leu-	Ergocornine	L-α-Aminobutyric acid	15
	Ergokryptine	5,5,5-Trifluoroleucine	21
		β-Hydroxyleucine	11

even be visualized as a process in which some physiological signals are elaborated in the form of oligopeptides affecting specific functions.

Cycloheximide does not affect the formation of the peptide part of ergopeptines.[47] The synthesis of the peptide portion of cyclol alkaloids is apparently a nonribosomal process. Genetic blocks in the biosynthesis of the peptide moiety or in that of the lysergyl moiety do not influence the biosynthesis of the remaining part of the molecule.[42]

Despite the very detailed knowledge of many aspects of ergot alkaloid biosynthesis, there are still a number of major blank spots which need to be filled in before we have a complete picture of the pathway. The most important aspects which still have to be worked out are the mechanism and enzymology of C-ring formation and the enzymology and formation of ring D and of the peptide moiety.

D. Ergopeptame

In a blind branch of biosynthesis of cyclol ergot alkaloids, lactam alkaloids called "ergopeptams" (Figure 16 and Table 4) with noncyclol structure and D-configuration of proline in the peptide moiety of the molecule are produced.[37,48,49]

E. Clavicipitic Acid

The formation of clavicipitic acid (Figure 17) is a major alternative to the biosynthesis of alkaloids in the utilization of DL-4-γ,γ-dimethylallyltryptophan (DMAT). There is ap-

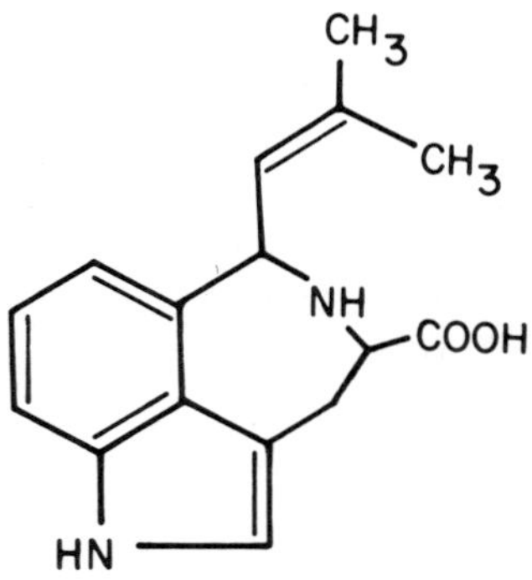

FIGURE 17. Clavicipitic acid.

parently no feedback control of clavicipitic acid formation by alkaloids since agroclavine does not inhibit DMAT oxidation significantly. DMAT oxidase, the enzyme catalyzing the conversion of DMAT into clavicipitic acid, and the alkaloid biosynthetic pathway may have developed to remove DMAT under different conditions.[52] This is reflected in the contrasting properties of the two pathways. DMAT is membrane bound, is produced during the exponential growth phase, and has a high pH optimum. The components of the alkaloid biosynthetic pathway are soluble, are produced during the phase of slow or no growth, and the pH optimum is near 7.

A cell-free system for the conversion of DMAT to clavicipitic acid was obtained.[52,53] Hydrogen peroxide was shown to be the product of oxygen reduction by coupling with peroxidase. The activity was independent of pH in the range 6 to 8 and increased to a maximum at pH 5, with a pK_{app} of 9.4. Enzyme-specific activity increased during the growth phase and then remained constant during the stationary phase. An intermediate compound could be envisioned which could either react further to close the C-ring and form chanoclavine-I or produce clavicipitic acid in a side reaction, which is not in the pathway of alkaloid biosynthesis.

F. Cell-Free Biosynthesis

Nearly all the knowledge concerning the formation and regulation of ergopeptines stems from feeding experiments with whole mycelium. Only the origin of the ergoline ring moiety from tryptophan, and dimethylallypyrophosphate was established at an enzymic level.[54-56] The cell-free biosynthesis of agroclavine, elymoclavine, chanoclavine-I, and chanoclavine-II from tryptophan, isopentenyl, pyrophosphate, and methionine was reported.[57] The cell-free synthesis of the ergot peptides from their constituent amino acids and D-lysergic acid has not yet been accomplished. At present, no enzyme preparation is available catalyzing the synthesis of the peptide moiety, e.g., a "cyclol synthetase complex". A method to obtain a suitable cell-free system for this purpose might be to use protoplasts actively synthesizing ergopeptines.

Maier et al.[58] prepared protoplasts of *C. purpurea,* which were able to synthesize *de novo* ergosine, ergotamine, and ergotoxine using ^{14}C-labeled amino acids as precursors. The incorporation rate of cyclol-specific amino acids may differ from strain to strain, the age and environmental conditions of the mycelium used for protoplast preparations, and the size of the internal pool of free amino acids.[59] However, it is not clear whether this pool is the cytoplasmic amino acid pool itself or whether it is located in a cell compartment such as the vacuole. Location of these pools would enable the determination of the intracellular locus of ergopeptine synthesis. The incorporation rate of the amino acids was remarkably increased if the incubation mixture was supplemented with D-lysergic acid.[58] The concentraiton of D-lysergic acid in the cells is a rate-limiting factor in ergot peptide synthesis.[59]

Tryptophan-^{14}C labeled also ergopeptines in protoplast suspension. The addition of methionine to protoplasts or mycelium led to a decrease of ergot peptide synthesis. A regulatory role of methionine in the control of this biosynthesis is supposed.

Characteristically, only floating vacuolated protoplasts were capable of synthesizing ergot peptides. It can be assumed that the synthetic capacity is a feature of this special physiological state.[60] As age-dependent vacuolation is accompanied by an increase in the oxidative metabolism of the cell, it is not surprising that ergot peptide synthesis is stimulated by shaking, in both the protoplast and mycelia suspensions.

In *Claviceps,* as in other genera, the fusion of protoplasts has a quite different effect whether it is intraspecific or interpsecific. The intraspecific fusion originates heterokaryons able to complement both nutritional requirements and alkaloid synthesis, whereas in the interspecific fusion aneuploids are formed which show nutritional requirement complementation only.[61]

IV. PHYSIOLOGY

Alkaloid biosynthesis is more a result of changes in the physiological state than an indication of a new physiological state. Genetic manipulation and knowledge of physiological controls are dominant tools of modern alkaloid production. While metabolic pathways taking part in the synthesis of ergot alkaloids have been studied in detail, physiological formation of alkaloids has been investigated only negligibly. The formation physiology does not consist only of culture growth and synthesis of the ergoline skeleton from corresponding precursors. The synthesis of alkaloids is correlated with distinct genotypes[62] and requires changes in the metabolic structure normally leading to propagation of biomass, a decrease in the intensity of culture growth, a decrease in the number of nuclei in the terminal cells of hyphae, and a change in differentiation.[63] The absolute values in physiological studies on ergot are greatly influenced by such factors as the age of the inoculum and the number of times the organism had been transferred prior to the preparation of the inoculum for the experimental cultures.

During the fermentation phase, degradation of alkaloids takes place but is overcompensated by alkaloid synthesis. When alkaloid-forming enzyme systems as well as the formation of degrading enzymes are inhibited by hydroxyurea (HU), some pre-existing enzymes still function to lower the alkaloid content to a certain level. Upon addition of HU together with cycloheximide (inhibitor of protein synthesis), the alkaloid content remains nearly unchanged.[64] This might provide a hint that some factor(s) for the action of alkaloid-degrading enzymes is affected by cycloheximide.

A. Fermentation

The era of saprophytic production of ergot alkaloids was inaugurated by the fundamental work of Professor Abe in Japan shortly after WW II. During this work the first culture medium was developed which is especially suited for saprophytic production of ergoline alkaloids.

Stationary cultures of *Claviceps* spp. develop as surface mycelial mats that cover the entire surface of the medium and are fully exposed to atmospheric oxygen, whereas the lower, partially anaerobic, surface is in direct contact with the growth medium. The kinetics of such growth is different and more complex than the kinetics in shaken cultures. There is often a heterogeneity of metabolic environment in fungal cultures. On the other hand, shake cultures of *Claviceps* spp. develop as a uniform mass or as pellets of mycelium, but are still not completely homogeneous. Shake cultures were not routinely studied, and the careful experiments that reveal relationships between growth and alkaloid production were not performed. As knowledge of the biochemistry of microbial growth becomes more complete, the data support the view that the relationship between growth and alkaloid formation is not quite as simple as originally thought by natural-product chemists.

Table 6
SOME CULTIVATION MEDIA FOR SAPROPHYTIC *CLAVICEPS* STRAINS PRODUCING ALKALOIDS[65]

Component (gL^{-1})	T2	SL	Pep 3	TG	TG/S	T25	T26
Sucrose	100	100	300	—	—	300	200
Glucose	—	—	—	100	100	—	—
Citric acid	—	—	—	10	—	15	—
Succinic acid	—	—	—	—	7.5	—	14
Asparagine	10	10	—	—	—	—	—
Peptone	—	—	10	—	—	—	—
Yeast extract	0.1	—	—	0.1	0.1	0.1	0.1
KH_2PO_4	0.25	0.25	0.5	0.5	0.5	0.5	0.5
$MgSO_4 \cdot 7H_2O$	0.25	0.25	0.5	0.3	0.3	0.5	0.5
KCl	0.12	—	—	—	—	0.12	0.12
$Ca(NO_3)_2 \cdot 4H_2O$	1.0	—	—	—	—	—	—
$FeSO_4 \cdot 7H_2O$	0.02	0.007	0.007	0.007	0.007	0.007	0.007
$ZnSO_4 \cdot 7H_2O$	0.015	0.006	0.006	0.006	0.006	0.006	0.006
NH_4OH (pH)	5.2	5.2	—	5.2	5.2	5.2	5.2
Agar	18	18	18	—	—	—	—
Water	T	D	T	D	T	T	T

Note: T = tap; D = distilled.

Although good growth of axenic *Claviceps* cultures may occur in many media, alkaloids may only be produced in a specific medium (Table 6). Individual *Claviceps* cultures are highly variable and the procedures used are nonuniform. Development of a medium that produces high yields of alkaloids is still empirical to a considerable degree. Distribution of precursors into various branch pathways may vary and may depend on the growth environment and the genotype of the organism. The flow of precursors can be increased in the desired direction by mutation or medium manipulation. The presence or absence of certain ions or of carbon and nitrogen sources can inhibit, activate, induce, and derepress certain enzymes, perturbing the normal channeling of key intermediates that support balanced growth.[66]

Ergot alkaloid fermentations are done in the temperature range of 20 to 30°C (favorable 23 to 26°C) and in a pH range of 4.6 to 6.3 (at best, 5.2 to 5.8). *Claviceps* spp. grow well in a medium containing inorganic salts supplemented with an organic carbon source and biotin. Nitrate, ammonia, or organic compounds can be used to supply nitrogen. Succinate, and probably other tricarboxylic acid cycle acids as well, are superior counter ions with ammonia as the nitrogen source because they are rapidly taken into the mycelium by diffusion at low pH values and metabolized. With increasing pH value, uptake decreases sharply as the acid dissociates and the acidity of the medium is balanced in a suitable range.[67] Phosphate (0.3 to 300 m*M*) supports extensive microbial growth but a concentration of inorganic phosphate of 10 m*M* and above exerts a depressive effect on the synthesis of many secondary metabolites belonging to different biosynthetic groups. Trace metals play a remarkable role in ergot alkaloid biosynthesis and may be involved in enzyme-coenzyme conditions.

High intracellular concentrations of tricarboxylic acid cycle acids may regulate metabolic activity,[215] and in media with a large excess of organic ammonium salts, the unused balance may assist by increasing the osmolarity and providing buffering capacity. High osmotic pressure obtained with sugar (100 to 300 g/ℓ) also appears to be beneficial.[68]

Nutrilites enter the cytoplasm either by osmosis (through the plasma membrane) or from the mitochondria. When the mitchondrial membrane contracts, pressure forces some solutes out of the mitochondrion into the cytoplasm. In either case only small molecules are involved. Osmosis controls most of the cytoplasmic nutrition.

In media containing high concentrations of sucrose, the uptake of oxygen by mitochondria is decreased.[69] Inhibition of mitochondrial respiration with DNA-linked substrates and with succinate occurs in hyperosmolar media containing sucrose, mannitol, NaCl, or KCl as the principal osmotic agent.[70] Under conditions of hyperosmolarity, different substrates show dissimilar degrees of inhibition of respiration.[71] This variation could be caused by changes in the selective permeability of the mitochondria under the condition of high sucrose concentration.

Mitochondrial NADH levels have been found to be lower in hyperosmolar than in normoosmolar media and uncoupling agents do not release the respiratory inhibition.[70] Therefore, a block in energy transfer does not appear to be the main cause of respiratory inhibition. The respiratory block in hyperosmolar media is situated before NAD with NAD-linked substrates and before cytochrome *b* with succinate as substrate. There may be an impediment to entry of substrates or a decreased accessibility of substrate to appropriate dehydrogenases under conditions of hyperosmolarity and this impediment is the principle cause of respiratory inhibition.

Under both parasitic and saprophytic conditions, measurements of $^{Q}O_2$ values have shown that the late growth phase, when alkaloids are normally produced, is associated with a decreasing oxygen demand from cells. However, an increasing redox potential in aging mycelium does not direct a shift from clavine to lysergic acid synthesis.

The literature provides conflicting reports on aeration. Very rigorous shaking of the culture and intensive aeration invariably leads to a low alkaloid concentration. The beneficial influence of aeration and the absolute requirement of molecular oxygen in the biosynthesis of alkaloids pointed to the possibility that oxygenases may participate in their biosynthesis. Ogunlana et al.[72] tested certain substances which are known to affect oxygenases for their influence on the alkaloid production in shake cultures of *Claviceps* strain SD/58. Sodium L-thyroxine produced inhibition; phenobarbital, hydrocortisone, and 3,4-benzpyrene caused marked increases. These findings lend strong support to the concept that oxygenases are involved in the biogenesis of the ergot alkaloid.

The effects of homogenizing of mycelium on the production of alkaloids differ considerably. Tabor and Vining[73] accelerated the production of alkaloids during stationary cultivation of *C. purpurea* PRL 1578 by homogenizing. Arcamone et al.[74] obtained reproducible yields with *C. paspali* F. Likewise, Pacifici et al.[75] eliminated the nonuniformity of the inoculum from agar slants in the isolate *C. paspali* C-60 by homogenizing. However, Gröger and Tyler[76] found a decreased concentration of total alkaloids after homogenization of the inoculum of *C. paspali* Li 189. Even when using the homogenized mycelium it is necessary to work out an inoculation procedure for every individual culture. The favorable effect of homogenizing on alkaloid production is associated with the characters of the strains used.

In the submerged conditions of the second fermentation stage, alkaloid production has been associated with pigmentation. The older the vegetation mycelium used for inoculation, the more intense is the production of a diffuse yellow-brown pigment.[77] Washing of the mycelium with distilled water resulted in a visible decrease of the pigmentation intensity in the subsequent stage and in a simultaneous increase in the production of total alkaloids. Washing of the mycelium before or after its homogenization did not exhibit a pronounced effect on alkaloid production. There is an inverse relation between the growth rate of the submerged culture and the intensity of alkaloid and pigment production.

The media composition, aeration, and agitation has to satisfy the biochemical requirements of *Claviceps* cultures. In the transition phase, very low ventilation is used.[85] This causes increased CO_2 partial pressure whereby it stimulates the glyoxylate cycle concomitant with inhibition of the mitochondrial function. In the production phase the culture shows an increasing oxygen demand according to which it is necessary to raise the aeration rate and the drive speed. Even at the high aeration-agitation rate the process is characterized by a relatively low oxygen transfer rate. The necessary oxygen transfer rate can be ensured by different methods of aeration-agitation.

Cultivation conditions suitable for the metabolism of primary substances are not suitable for alkaloid metabolism.[78] Submerged *Claviceps* cultures form alkaloids after using up the phosphate in the medium and accumulating polyols, lipids, polyphosphates, nonstructural carbophosphates, and inducers and precursors of alkaloids. Successful course of the alkaloid synthesis requires the presence of higher citrate level or some other intermediates in the tricarboxylic acid cycle in the cell. Cell differentiation and the onset of alkaloid synthesis is governed both by phosphate limitation[79,80] and citrate utilization. The synthesis of alkaloids is unfavorably influenced by an increased phosphate concentration in a fermentation medium.[81,82] High production strains are usually tolerant to higher phosphate concentrations.[83] The results obtained so far indicate that phosphate plays a negative role in the induction of enzymes catalyzing alkaloid synthesis.[84] After more specific studies, the opinion was reached that high phosphate levels limit tryptophan synthesis. Optimization of the fermentation process by adjusting phosphate dosage would require immobilization of nongrowing cells from the production phase on a suitable carrier, and inflow of the medium which would contain phosphate at a concentration of 0.030 to 0.035 g/ℓ.[85]

The high amount of phosphate piles up in the cells as polymetaphosphate.[86] Either orthophosphate or polymetaphosphate can inhibit glucose-6-P-dehydrogenase, the key enzyme of the hexose monophosphate cycle. Good proof was given of an inverse proportionality between the rate of alkaloid synthesis and the activity of the tricarboxylic acid cycle.[87] The levels of the tricarboxylic acid cycle enzymes decreased in favor of glyoxylate enzymes. The addition of α-ketoglutarate caused a twofold increase in the alkaloid level — the same as if succinate were added. Phosphofructokinase exhibited very low activity in *Claviceps* cultures when forming alkaloids.[88] This was found to be indicative of an active hexose monophosphate cycle.

Acceptance of citrate from the medium can be temporarily increased by increasing the concentration of phosphate in the medium. Cells cultivated in high phosphate medium show decreased levels of acid and alkaloid phosphatase, and the occurrence of novel isoenzymes of alkaline phosphatases in the alkaloid-producing phase which are repressed in high phosphate conditions.[89]

In submerged cultures of *C. purpurea,* citrate utilization was stopped when the pool phosphate concentration decreased to 0.44 to 0.55 mmol/g of mycelial protein.[90] The absence of citrate or another tricarboxylic acid cycle intermediate in the medium induced a specific complex of differentiation abberations which resulted in a 50% decrease of biomass, cessation of alkaloid synthesis (agro-, elymo-, chano-clavine), and medium acidification by lactate. The cells of cultures lacking citrate were enlarged and characterized by a central large vacuole and a gradual disappearance of citrate synthase and malate dehydrogenase. Glucose-6-P-dehydrogenase was not detected. The activities of both glycolytic enzymes increased to values higher by one or two orders of magnitude than in the controls. In the cultures lacking citrate, the lactate dehydrogenase exhausted pyruvate to reoxide NADH. Depletion of citrate precursors, which were normally formed from pyruvate, led to a shift towards glycolysis.

Three types of effects of citrate were observed:[86] (1) increase of the alkaloid level, (2) increasing succinate pool in the cell, (3) a sudden increase in lysine synthesis at extreme levels of citrate. Citrate is an intermediate which has an important role in metabolic regulation. It inhibits phosphofructokinase and stimulates glucose-6-P-dehydrogenase. Thus, it enhances the shift to the hexose monophosphate route. Increasing amounts of citrate may be converted via α-ketoglutarate into the products of the glyoxylate cycle. This is shown in the accumulation of cell pool succinate. Further amounts of α-ketoglutarate so formed can be eliminated in the fungal path to lysine biosynthesis.

We assume that the inability of submerged *Claviceps* cultures to grow on hexoses in the absence of the tricarboxylic acid cycle intermediates comes from the parasitic way of life of this pyrenomycete, where the level of citrate and malate in the phlem sap of the host plant is high. Therefore, in parasitic strains the impairment of mentioned regulatory relationships has not been excluded during the *Claviceps* evolution as a selection disadvantage.

Claviceps spp. cultures generally produce a variety of alkaloids with related chemical structure. The shifting of the component ratio, the repression of the synthesis of undesirable components, or the promotion of the yields of minor components could be enhanced by the knowledge of the regulation of the biosynthetic process.

Ergot alkaloids are produced under conditions of substrate limitation other than carbon from key intermediates in basic metabolism by pathways differing from basic metabolism. Many enzymes and coenzymes of basic metabolism and alkaloid synthesis are common. A reaction in which enzymes do not participate must not be neglected. In submerged cultures, alkaloid synthesis occurs under conditions of decreased proliferation and clearly defined degradation and resynthesis of nucleic acids and proteins.[47,91,92] The growing mycelium of *C. purpurea* producing clavine alkaloids exhibits protein turnover, which is substantially reduced or even absent in the stationary growth phase.[93] This pattern is different from bacteria, where protein turnover increases in the stationary growth phase.[94] In following the physiology of the formation of ergot alkaloids, the metabolic pathways of histidine and tryptophan are worth studying as they involve not only repression and feedback inhibition, but also cross-pathway regulation.[91,92]

Three phases of culture growth are observed in the ergotamine fermentation with *C. purpurea* strain 275 FI in 800 L fermentor on a standard sucrose-ammonium citrate.[83] The first growth phase ends about 70 hr, when organic phosphate disappears from the medium and is characterized by a high $^{Q}O_2$ value, by exponential growth, and by active ammonium nitrogen uptake. The second linear growth phase (70 to 180 hr) corresponds to a transition phase between the time of exhaustion of phosphate and that of nitrogen. In the typical ergotamine fermentation, free glucose originating from sucrose in the early stage of growth is exhausted before ammonium nitrogen, but other carbon sources and citric acid are still available. The third growth phase (180 to 280 hr) appears to be a kind of ''maintenance'' phase during which the dry weight increases at a slower rate with concomitant moderate uptake of carbohydrate. Approximately one half of the total lysergic acid derivatives are synthesized during the third phase.

An accumulation of ergot alkaloids of *C. purpurea* occurs during a prolonged transition phase that is characterized by the accumulation of nitrogenous, but not carbohydrate, storage material.[78] This phase falls between the balanced and storage phase, where magnesium and phosphorus were limiting nutrients. Mycelial proliferation continues, but cellular composition changes from that of the balanced growth phase. During growth in some media that do not support alkaloid production, cells pass from balanced into storage phase and store carbohydrates, polyols, and lipids.[86,88] During logarithmic growth, a growth-linked overside mechanism represses the synthesis of the enzymes responsible for the formation of ergot alkaloid by *C. purpurea.*[95]

Tryptophan is not only a precursor of ergot alkaloids, but is also a factor in the induction and derepression of enzymes catalyzing alkaloid formation[84,96,97] (Figure 18). Exogenous tryptophan removes inhibition of alkaloid production by phosphate consisting of derepressed tryptophan synthesis.[81] The first enzyme of the tryptophan pathway, i.e., anthranilate synthetase, of the *Claviceps* strain SD 58 was inhibited in the presence of tryptophan, chanoclavine, elymoclavine, indoleacrylic acid, and prephenic acid, and was slightly activated in the presence of histidine.[98] Only one tryptophan cell pool was detected in *Claviceps* cultures.[98] The decrease in the content of the cell protein tryptophan has a favorable effect on the biosynthesis of alkaloids.[91,100] Dimethylallyltryptophan (DMAT) synthetase, the first unique enzyme in alkaloid synthesis, is induced by tryptophan.[84] An inductive effect similar to that observed in alkaloid synthesis is seen in the methionine stimulation of cephalosporine C biosynthesis by *Cephalosporium acremonium.*

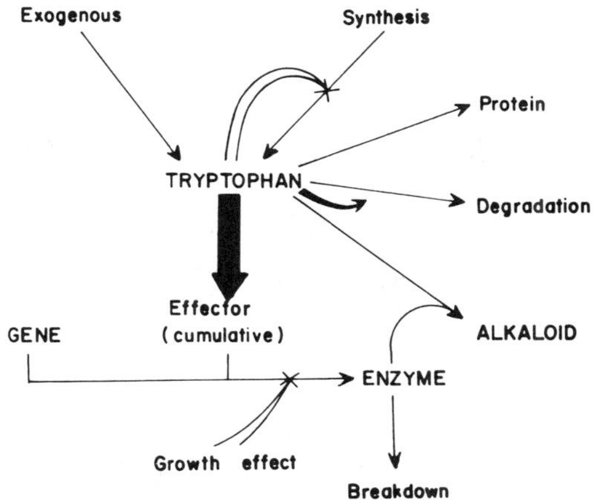

FIGURE 18. Different roles of tryptophan in axenic *Claviceps* cultures.

The regulatory role of tryptophan can be mimicked by thiotryptophan, which induces DMAT synthetase. During rapid growth, tryptophan normally accumulates in the mycelium, but the induction of enzymes by accumulated amino acid does not occur until phosphate depletion disrupts growth, thus culminating in ergoline alkaloid production. At this time, some growth-linked override regulatory mechanism ensures that the transcription of genes involved in ergoline biosynthesis is no longer prevented. A high growth rate may alter the amount of some regulatory molecules (e.g., highly phosphorylated nucleotide) that participates in transcription.

Transcription of genes involved in alkaloid biosynthesis may not begin even in the presence of inducing levels of tryptophan, which was probably removed by the repressor. It is possible that when exhaustion of nutrients (such as phosphate) reduces the growth rate and adjusts the intracellular concentration of a regulatory molecule, the overriding regulatory mechanism, which is growth rate-linked and represses induction of ergoline biosynthesis, is lifted, and genes specific for ergoline synthesis are expressed. An alteration of RNA polymerase may also be one of the pleiotropic controls that govern the expression of genes specific for alkaloid synthesis.

Addition of hydroxyurea (HU), which selectively inhibits DNA synthesis by affecting ribonucleotide diphosphate reductase,[101] affects the enzyme-forming system and interrupts the metabolic shunts towards ergot alkaloid formation.[93]

B. Carbohydrate Metabolism

The cultures with dominant elymoclavine yields possess an intensive oxidative metabolism, an efficient supply of glycolytic intermediates to the tricarboxylic acid cycle, and a higher activity of the hexose monophosphate pathway.[102] The oxidative metabolism and a sufficient NADPH supply is involved in hydroxylation of agroclavine to elymoclavine.

1. Glycolytic Pathway

In a submerged *Claviceps purpurea* culture producing clavine alkaloids, phosphofructokinase (PFK), a key regulatory enzyme in the glycolytic pathway, exhibits hyperbolic kinetics which is not typical of this enzyme.[103] Inhibition of PFK by glucose or some of its related metabolites points to a feed-forward regulation of PFK and to a possible new regulation model of the glycolytic pathway (Figure 19). The mechanism of feed-forward regulation was proven with arginine-tRNA synthetase inhibited by arginine succinate, and was thought to be exceptional. The PFK feed-forward regulation in *C. purpurea* can reflect the fact that the regulatory mechanisms in the original parasitic *Claviceps* cultures dependent on the host

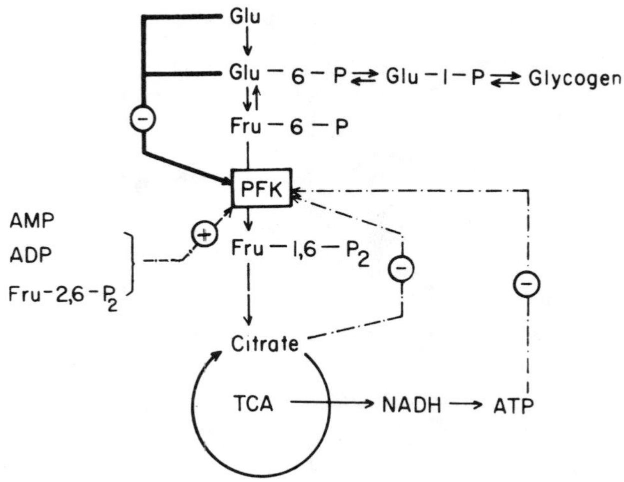

FIGURE 19. Newly proposed feed-forward regulation in the glycolytic pathway of *Claviceps purpurea* producing clavine alkaloids.[103] — Feed-forward regulation; — · — classical feedback regulatory mechanism; ⊕ stimulation; ⊖ inhibition.

plant for nutrient supply weakened during evolution and the fungus became "lax" in this respect. This is also indicated by the finding that the *C. purpurea* PFK exhibits hyperbolic kinetics and is not stimulated by fructose 2,6-bisphosphate. In resting cells the absence of oxygen would prevent the formation of the great proportion of ATP formed in respiration. As a consequence, various regulatory mechanisms would be switched on,[104] producing the increase in the rate of glycolysis.

2. Hexose Monophosphate Pathway

The hexose monophosphate pathway plays a key role in the formation of NADPH, pentoses, and erythrose necessary for biosynthetic reactions. A crucial function in the regulation of this pathway has been attributed to glucose-6-phosphate-dehydrogenase (GPD), while a secondary role has been ascribed to its partner enzyme, 6-phosphogluconate dehydrogenase (decarboxylating) (PGD). Both the induction of enzymes and the modulation of enzyme activity at the substrate level are considered to be important in regulating the capacity of these dehydrogenases.

NADPH, ribose 5-phosphate, and erythose 4-phosphate, all known intermediates of the hexose monophosphate pathway, are needed for the biosynthesis of several amino acids. Therefore, for cells growing in a medium supplemented with amino acids, a great decrease of the cellular NADPH requirement can be expected and, actually, the activities of GPD and PGD decrease. Deregulation of the hexose monophosphate pathway generates from glucose excess intracellular ribose, which is the precursor for nucleosides. Glucose provides acetate, propionate, and NADPH for the successive reduction steps.

3. Tricarboxylic Acid Cycle

In the beginning of submerged cultivation of the *Claviceps* strain with intense cell proliferation, the tricarboxylic acid cycle is the predominant amphibolic pathway. It is the most effective component in carbohydrate degradation. The activity of its key enzymes is one to two orders of magnitude higher than that of the glycolytic and hexose monophosphate enzymes. With citrate synthetase, a negligible activity has been found in productive *Claviceps* cultures and three- to fivefold higher activity in a strain with a very low alkaloid content. Accumulation of citrate suppresses the citrate synthetase activity and activates acetyl-CoA carboxylase.[105] Under these conditions, a greater amount of malonyl-CoA is formed and a greater number of condensation reactions take place. During alkaloid synthesis, acetyl-CoA enters the tricarboxylic acid cycle to a limited degree and takes an alternative path to the synthesis of fatty acids. The lipid content in the mycelium increases almost parallel to the activity of acetyl-CoA carboxylase and the total alkaloid level.[91,106]

4. Glyoxylate Cycle

In the alkaloid phase, the cultures gain energy and intermediates for synthetic processes chiefly through the glyoxylate cycle.[87,107] During the alkaloid phase, the glyoxylate cycle simultaneously enables the introduction of pyruvate into cells. 2-Ketoglutaric, isocitric, citric, and fumaric acids stimulate the synthesis of alkaloids in cells with an intense glyoxylate cycle and decreasing activity of isocitrate dehydrogenase.[107] The glyoxylate cycle contains some reactions common with the tricarboxylic acid cycle. In eukaryon cells the two cycles are physically separated. Most enzymes of the glyoxylate cycle are located in organelles and are termed ''glyoxysomes''. Mitochondria are the components of the enzymes of the tricarboxylic acid cycle. Possible participation of glyoxysomes and mitochondria in the regulation of alkaloid biosynthesis cannot be excluded.

C. Enzymic Pattern

Ergot alkaloids are produced from a few key intermediates of primary metabolism by a consecutive series of enzyme-catalyzed reactions along pathways which branch from the primary network at a relatively small number of points. The enzymes associated with ergot alkaloid synthesis can be divided into four classes:

1. Those of primary metabolism that yield the precursors of alkaloids
2. Those that are specific for the synthesis of alkaloids
3. Those producing energy or cofactors or inserting various functional groups into the biosynthetic intermediates
4. Those that further metabolize the produced alkaloids

Enzymes that carry out primary metabolic processes have evolved with strict substrate specificity because their product is essential to the growth of the organism. However, enzymes involved in the biosynthesis of alkaloids have not always been selected for rigid specificity because alkaloids are not essential for the growth of the producing organism. The synthesis of alkaloids is carried out by specific pathways involving enzymes governed by genes that are not involved in growth.[108]

Enzymology of ergot alkaloid biosynthesis in *Claviceps* spp. unravels some very interesting phenomena. Even though the formation of secondary enzymes is generally repressed during logarithmic growth and is derepressed during suboptimal or stationary growth, and even though their synthesis is controlled by regulatory mechanisms, initiation of alkaloid biosynthesis frequently coincides with the exhaustion of some major ingredient of the growth medium and cessation of growth. Synthesis of alkaloids by *C. purpurea* begins after all the phosphate has been utilized.[95,109] The main enzymes of tryptophan biosynthesis are of higher activity than in nonproducers and show another profile in activity during alkaloid fermentation.[110]

The formation of ergot alkaloids occurs by multistep processes catalyzed by a multienzyme complex that usually is produced only during a certain growth phase. Thus, some of the alkaloid-synthesizing enzymes are induced during a short period at the end of logarithmic growth and the onset of the stationary phase. They are also unstable and are in many cases inhibited by the products of the reaction they catalyze. As a result, most of the published curves that show the relationship between enzyme activity, culture growth, and alkaloid production could be misleading because the enzyme preparation used for plotting such curves was not extensively purified to remove inhibitory enzymic reaction products.

The efficiency of synthesis is attributable to genetically controlled enzymic levels and special enzyme properties. Intracellular compartmentalization of enzymes, substrate, and cofactors makes the biosynthetic processes highly efficient. Multienzyme complexes with catalytic subunits arranged to catalyze successive steps of synthesis are known to be involved in the biosynthesis of secondary metabolites. The catalytic efficiency of these complexes is probably higher in the intact cell than in cell-free extracts because the intracellular intermediates are largely enzyme-bound during the biosynthetic process. Toxic enzyme-bound intermediates may not inhibit cellular functions because they are never free in the cytoplasm. In the cell-free extract, however, this delicate order is disturbed with the ensuing dilution of the desired enzymic activities.

1. Dimethylallyltryptophan Synthetase

The dimeric enzyme dimethylallyltryptophan synthetase (dimethylallylpyrophosphate:L-tryptophan dimethylallyltransferase) has been purified from *Claviceps* strain SD 58.[111] It contains two similar subunits of 34,000 mol wt. It is specific for dimethylallylpyrophosphate. Analogs of tryptophan-carrying methyl groups in the benzene ring showed decreased substrate activity. The closer the methyl group was to the reactive 4 position, the lower the activity. Dimethylallyltryptophan (DMAT) is converted into tetracyclic ergolines as an intact unit, i.e., without cleavage of the bond between the amino nitrogen and the *N*-methyl group.[112] The deregulated DMAT synthetase has K_m values of 7.2 μM for dimethylallylpyrophosphate and 8.8 μM tryptophan. DMAT synthetase catalyzes the prenylation of tryptophan by DMAPP. The prenyltransferase of DMAT synthesis as well as the prenyltransferase (avian and pig) of farsenylpyrophosphate synthesis all have similar molecular weights and subunit composition.[113]

A concentration of 20 mM Ca^{2+} completely deregulates the DMAT synthetase while 30 mM Ca^{2+} ion is inhibitory. Since calcium is not essential for catalytic activity, it most likely does not have a catalytic function and is probably involved only in regulation. In its natural habitat as a plant pathogen, *Claviceps* is in an environment of high calcium content. Normal levels of calcium are approximately 125 μmol/g of dry matter in young and herbacious plant tissue,[114] which is about six times the concentration needed for complete deregulation of the enzyme in vitro. Under in vivo conditions, the complete deregulation of DMAT synthetase is thus possible, allowing for a maximum production of alkaloids under natural conditions.

DMAT synthetase is inhibited by the end products of the pathway in *Claviceps* strain SD 58, agroclavine and elymoclavine.[115,116] The enzyme from a peptide alkaloid-producing strain was not inhibited by clavine alkaloids. The inhibition is not competitive with either substrate. The concentrations of agroclavine and elymoclavine which inhibit the enzyme are in the range of concentrations in the cultures. The enzyme is not synthesized until the period of growth is over. Enzyme synthesis immediately precedes biosynthesis of the alkaloids.

2. Dimethylallyltryptophan N-Methyltransferase

N-Methyl-4-(γ,γ-dimethylallyl)tryptophan, but not its decarboxylation product, is an intact precursor of the tetracyclic ergoline elymoclavine.[112] An *N*-methyltransferase catalyzes the transfer of the *S*-methyl group of *S*-adenosylmethionine to the amino group of 4-(γ,γ-dimethylallyl)tryptophan. This methylation is the second pathway-specific step in ergoline biosynthesis. Pyridoxal phosphate cannot be involved as cofactor in the decarboxylation and C-ring closure steps in this biosynthesis. This follows from the fact that pyridoxal phosphate catalysis requires a complanar arrangement of the imine double bond and the pyridine ring which is no longer possible in an iminium ion derived from an *N*-methylated amino acid or amine, even if such an iminium ion can be formed.

The detection of DMAT *N*-methyltransferase activity[112] brings to four the number of enzymes involved in ergoline biosynthesis which have been purified from or characterized in cell-free extracts of *Claviceps*. The other three are DMAT synthetase,[117] chanoclavine-I cyclase,[118] and agroclavine hydroxylase.[119] The 8-hydroxylation of agroclavine and elymoclavine and the conversion of dimethylallyltryptophan to clavicipitic acid (Figure 17) probably occurs by similar mechanisms.

3. Chanoclavine Cyclase

Enzyme systems catalyzing ring-D cyclization have been isolated in several laboratories, but they are difficult to purify. Ogunlana et al.[120] obtained 20% conversion of chanoclavine-I to elymoclavine. Besides ATP, magnesium ions, NADPH, and molecular oxygen were required. Agroclavine was not produced and was poorly converted into elymoclavine. The investigators proposed a pathway from chanoclavine I to elymoclavine without the intermediacy of agroclavine. Gröger and Sajdl[55] and Erge et al.[118] obtained a cell-free system which produced agroclavine but not elymoclavine in a 10 to 30% yield. The cell-free system of Heinstein et al.[121] required NADP but not ATP and catalyzed the formation of both agroclavine and elymoclavine. The addition of FAD to column-fractionated preparations stimulated activity. The requirement for FAD and NADP(H) indicated that an oxidation-reduction reaction is involved in the conversion of chanoclavine-I to agroclavine. Formation of chanoclavine-I aldehyde and/or hydride ion transfer in the isomerization reaction are oxidation-reduction reactions that have been proposed as part of the chanoclavine-I cyclase mechanism.

4. Agroclavine Hydroxylase

A step in the pathway of ergot alkaloid biosynthesis is the hydroxylation of agroclavine to elymoclavine. The hydroxyl oxygen did not come from solvent and may be concluded to be derived from molecular oxygen. Alkaloid-producing strains of *Claviceps* spp. carried out the conversion of agroclavine to elymoclavine, whereas nonproducing strains did not.

Hsu and Anderson[119,122] obtained conversion of agroclavine to elymoclavine with 60 to 80% ammonium sulfate fraction from *C. purpurea* PRL 1980. For activity, NADPH or liver concentrate (mixture of cofactors) was required. Although few microorganisms can hydroxylate the C(17)-methyl,[30] a microsomal system for liver can convert agroclavine to elymoclavine and noragroclavine.[123] This system involves cytochrome P-450.

The requirement for NADPH and lack of inhibition by EDTA and cyanide observed for the microsomal agroclavine hydroxylase are properties expected of a cytochrome P-450 monooxygenase. In contrast to the microsomal agroclavine hydroxylase, the ammonium sulfate fraction activity was inhibited by EDTA and cyanide.[122] The agroclavine hydroxylase activity in the ammonium sulfate fractions thus appears to be different from the microsomal agroclavine hydroxylase activity. The microsomal enzyme appears to be the enzyme in the main pathway of ergot alkaloid biosynthesis.[123a] The close relationship between enzyme activity and alkaloid production, the specificity for agroclavine as substrate, and the formation of elymoclavine as the only product indicate that the microsomal agroclavine hydroxylase is a component of the ergot alkaloid biosynthetic pathway. Cytochrome P-450 may be involved in other biosynthetic steps in ergot alkaloid biosynthesis, in ergosterol biosynthesis, and in the synthesis of ergot pigments.

5. Cytochrome P-450

There are two main pathways of electron transport: one from NADPH and the other from NADH. The highlights of the topics are concerned with cytochrome P-450 and its related oxygenase system. Cytochrome P-450 is the hemoprotein part of a mixed function oxidase system, which is found in a great variety of organisms. In microbial cells it is produced during a relatively short growth period after which it disappears rapidly.[124] Cytochrome P-

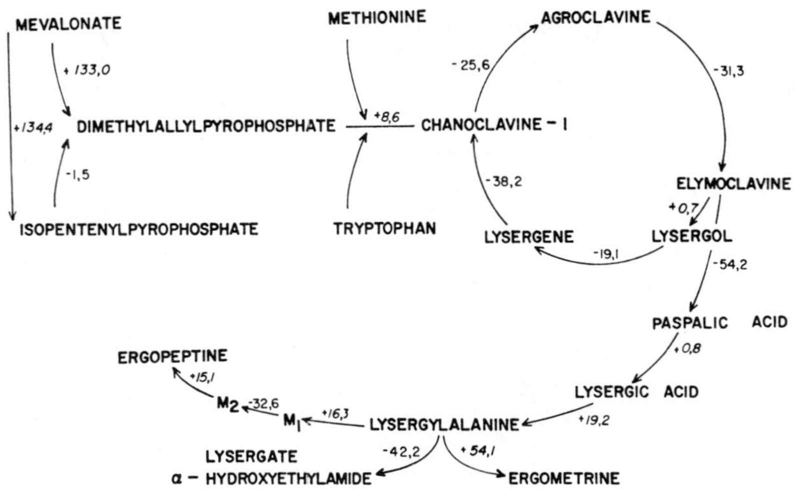

FIGURE 20. Mutual relationships and balance of the Gibbs free energy (free enthalpy) in the ergot alkaloid formation.[127] The numbers refer to the free energy charge $\Delta G^{o'}$ in kilocalories per mole reactants at 24°C, pH 7.0

450 monooxygenase binds substrate and molecular oxygen and interacts with reductases in a two-step sequence to activate the oxygen. Ultimately, one oxygen atom is incorporated into the substrate and the second atom is reduced to water. Isoenzymes of cytochrome P-450 are increasing in number since new ones are found every year. These isoenzymes play different roles in the incorporation of oxygen to or demethylation of physiological or xenobiotic substances.

Ambike et al.[125] demonstrated the presence of a cytochrome P-450 monooxygenase in the microsomal fraction of a *C. purpurea* strain that produces clavine-type alkaloids. Time course studies showed a direct correlation between alkaloid production and P-450 monooxygenase levels, thereby indicating an in vivo function of the enzyme. The P-450 monooxygenase was not ihhibited by cyanide. The enzyme was shown to be induced twofold by Pb. The increase in alkaloid production paralleled P-450 induction. Tryptophan did not increase cytochrome P-450 levels, but was synergistic in this regard with Pb. Following removal of cytochrome b_5 by steapsin (a lipase), the conversion of cytochrome P-450 to cytochrome P-420 was effected by snake venom protease, sodium cholate, or oxygen.[126] The reconversion of P-420 could be accomplished in quantitative yield by treatment with 10% glycerol, which also prevents the P-450 to P-420 transformation by steapsin. Cytochrome P-450-linked and cytochrome b_5-related electron pathways have been found to be closely interrelated. Studies on the mode of location of these electron transporting components in the microsomal membrane are recent projects of research.

D. Thermodynamics

The Gibbs free energy balance in the formation of ergot alkaloids[127] indicates that one of the key reactions in the biogenesis of ergot alkaloids is the formation of dimethylallylpyrophosphate (Figure 20). This energetically demanding reaction precedes enzyme allylation of tryptophan in position C(4) which is poorly accessible chemically. In contrast, the subsequent cycle of clavine alkaloids is part of the endogenic metabolism releasing the energy necessary for maintaining the integrity and viability of the cells under conditions of limited growth or even under conditions in the absence of growth and death of the mycelium. The energy of the clavine cycle is reminiscent of the ''energy of maintenance''[128] taking part in the resynthesis of the macromolecular components of the cell. This assumption is supported by the free energy balance of the formation of clavine alkaloids, by the low activity of the citrate cycle during the synthesis of clavines,[129] and by data indicating that the energy required for the turnover of the macromolecules is derived primarily from sources which the cell does not drain for its growth.[130] Considering that the evolution pressure results in types of cells marked by especially economical use of nutrients, it is possible to consider the positive role of the clavine cycle in the evolution of producers of clavine alkaloids. The manner of transformation of the biological energy of the clavine cycle is common for various taxonomically different organisms.[131] The formation of clavine alkaloids is accompanied by a decrease in the free energy with a tendency to achieve higher entropy. The molecules of derivatives of lysergic acid represent a more regular arrangement than clavine molecules. In the formation of derivatives of lysergic acid the free energy increases and entropy decreases. The free energy balance in the formation of ergot alkaloids contributes, among other things, to the explanation of the frequent occurrence of clavines and of the α-hydroxyethylamide of lysergic acid and of the relatively rare occurrence of ergometrine and ergotamine in ergot cultures. An important role in the occurrence of clavine alkaloids in nature is also played by other factors, e.g., the mechanisms of *C. paspali,* which protects sclerotium from deterioration through soil microorganisms, insects, and root excretions of plant.[132] *C. purpurea* lacks these protective mechanisms in its sclerotium; its sclerotium survives in the soil for approximately 6 months.

E. Differentiation

As already mentioned, the synthesis of alkaloids requires a change in differentiation. Differentiation is considered to be a result of different expression of the genome under various conditions and at different times.[133,134] It is the stable development of an altered structure and function.[135] This criterion emphasizes the irreversibility of the process and excludes transient changes which are referred to as ''adaptation''.

1. Nucleus and its Adjuncts

The nucleus, the mitochondria, and the endoplasmic reticulum comprise a single system separated from the cytoplasm.[136,137] The three components of this system are, respectively, ingestive, metabolic, and excretory. The nucleus and its adjuncts, i.e., mitochondria and endoplasmic reticulum, are oxidative. The cytoplasm is reductive. Open communication between the nucleus and the cytoplasm will destroy both.

The nucleus and the cytoplasm are stable, but are chemically incompatible hereditary systems. Mitochondria (Figure 21), the energy-generating organelles of the cell, have their own genetic system. The organization of their genes and even their version of the ''universal'' genetic code are different from those of nucleus and bacteria. Mitochondria have free access to both the nucleus and the external environment of the cell. The mitochondria are primarily hydrostatic suction pumps that concentrate the outer fluid and transfer it to the nuclear vacuole, thus exposing the genes to a magnified sample of the environment. The endoplasmic reticulum is a tube that continues the processing and finally carries waste from the nucleus to the outside of the cell by an open pipeline. If the flow through the endoplasmic reticulum is sufficiently rapid, no visible poison appears in any part of the cell.

2. Polarization Gradient

Cellular morphogenesis is accompanied by changes in the translational system. For *Mucor racemosus* most pronounced are the methylation and phosphorylation of elongation factor and ribosomal protein, respectively.[138] Morphological differentiation is preceded by synthesis of membranes, i.e., proteins and lipids. Protein synthesis is quantitatively regulated during morphological changes of *C. purpurea.*[87] For initiation of differentiation, macromolecular turnover with minimum requirements on the synthesis of new macromolecules is important. The metabolism of the early sequences of fungal differentiation in general is characterized

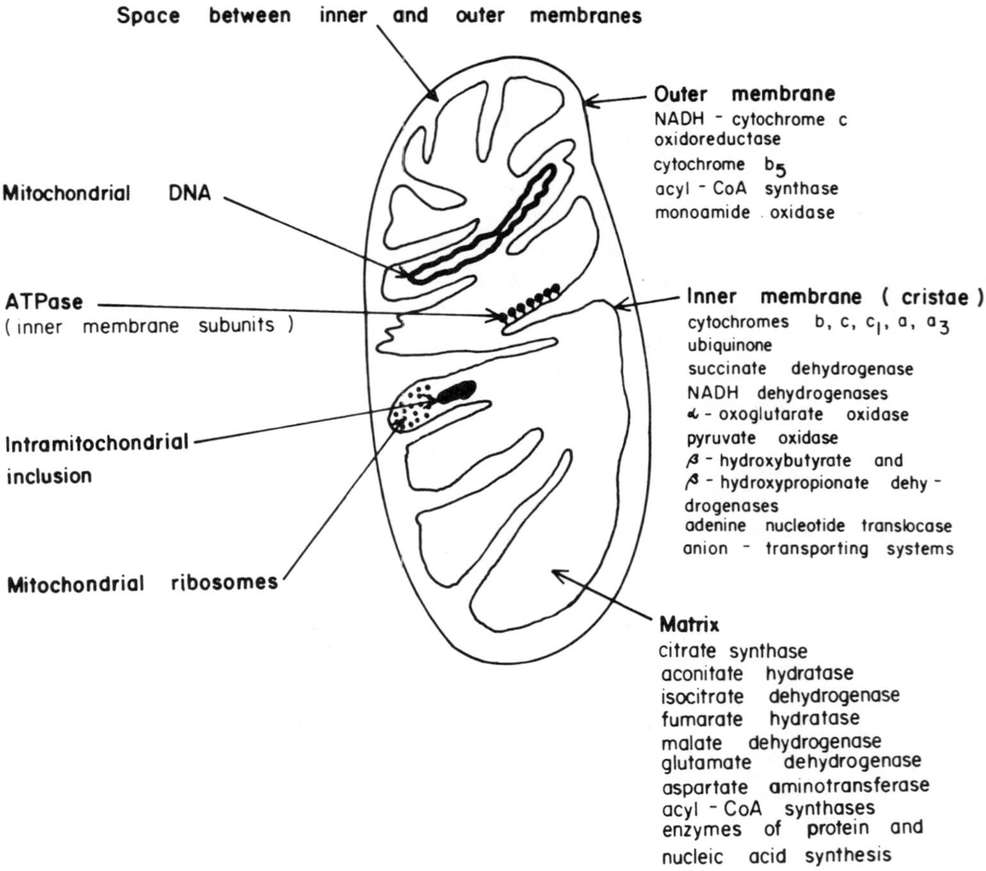

FIGURE 21. Structure and enzyme system of the mitochondrion.

by an increase in the relative importance of the hexose monophosphate pathway and that of the glyoxylate cycle; NADP-dependent functions predominate and there is a decrease in oxygen consumption.[139]

At the beginning of their differentiation, vegetative fungal cells form a polarization gradient in the cytoplasm. Respiration plays an important role in maintaining polarized growth. Hyphae growing under aerobic conditions require active mitochondria if they are to dominate in culture over cells with yeast growth. In vegetatively polarized hyphae, apical vesicles move through the cytoplasm along an electrochemical gradient produced by the oxygen. These vesicles contain precursors of the cell wall and hydrolyzing or synthesizing enzymes. The amitochondrial peak of vegetative hyphae is placed with reduction ability, provided glycolysis occurs during catabolite repression. This marked reduction ability is a result of sulfhydryl groups, which are very numerous at the hyphae end.[140,141] Turian[142] feels that glycolytic activity has a "vegetating" role, as conversion of the metabolism to the oxidative one induces conidia formation.

3. *Morphogenesis*

The first attempts to clarify the morphogenesis of submerged cultures of *C. purpurea* synthesizing ergopeptines revealed the thickening of filamentous hyphae and their disintegration into pellets of ovoid cells.[143] Such sclerotized hyphae and plectenchymatic pellets might be represented by both ovoid and elongated chlamydospores.[144] We observed the tissue-like clusters of extended hyphal cells and a yeast-like budding of these cells which resulted in the formation of asexual spores-blastospores.[145] These deciduous spores were borne by extended hyphal cells and retained the same ultrastructure of cell organelles. Both the extended hyphae and the blastospores resembled the cells of ergot sclerotial tissue. The fine cytoplasmic structure and the rare occurrence of vacuoles and organelles of the endoplasmic reticulum in blastospores corresponds to the ultrastructure of parasitic spores in the honeydew.[146]

The hyphal walls of submerged cultures of *C. purpurea* and *C. fusiformis* contain glucose, mannose, and galactose as the main neutral sugars.[147] The principal sugar accounting for more than half of this particular fraction is glucose. The amino sugar fraction (predominantly glucosamine) varies in different strains (e.g., in the ergotoxine producer Pepty 695) between 20 to 37%. The main fatty acids are lauric acid, myristic acid, and palmitic acid. *Claviceps* cell wall material has 5 to 10% protein. Protein hydrolysis gives 15 amino acids, predominantly threonine, serine, aspartic, and glutamic acid.

During the alkaloid-production phase the cells do not exhibit any special organelles. Although the chlamydospores and blastospores were proved to be present only in alkaloid-producing cultures, the appearance of numerous lipid droplets in these spores was indicative of culture aging rather than of alkaloid synthesis. The affinity of the lipid droplets toward the vacuole tonoplast seems likewise to be a general phenomenon.[148] The vacuolar bodies have been proved to contain cytochemically detectable polysaccharides which obviously represent the activity of the mannan-synthase complex.[149] The next step in the study of the relations between the production of alkaloids and the ultrastructure of *C. purpurea* is thus the detailed investigation of the cytochemical localization of the enzymic complexes ensuring the alkaloid synthesis. This concept is supported by the recent study of the ergot parasitic cycle,[150] which showed that alkaloid synthesis was not accompanied by any significant change in the pattern of basic cellular components in the sclerotized host tissue, e.g., a change in the level of carbohydrates, lipids, nucleic acids, amino acids, phosphate, or malate.

Sporulation and ergopeptine formation are initiated at the end of logarithmic growth when certain nutrients in the growth medium are depleted. The fact that alkaloid synthesis and sporulation are initiated at the end of growth makes analysis of the specific genes very difficult. Many genes are released from growth-linked repression at the end of the growth phase, and it is difficult to differentiate between these derepressed gene functions and those that are specific for growth, sporulation, or alkaloid formation.

With *C. purpurea,* the formation of conidia is inversely dependent on the synthesis of alkaloids for those saprophytic strains which partially retain parasitic development, i.e., differentiation of the sphacelial phase to the conidial and/or sclerotial phase. In contrast, in those strains for which the conidial phase, the phase of vegetative growth, and the alkaloid phase are not clearly separated, conidia formation occurs simultaneously with alkaloid synthesis.[151] These strains are, however, quite sensitive to external conditions. Physiological conditions necessary for conidiation and germination of conidia are different from conditions controlling alkaloid synthesis.[152] The conidiation process is affected by similar factors as the formation of alkaloids. In *Aspergillus nidulans* cultures, conidia formation is controlled by 300 to 800 genes.

The character of fat hyphae and sclerotia-like cells of submerged mycelium of *C. purpurea* is reminiscent of the plectenchymatic structure of production sclerotia of parasitic cultures. The sclerotia-like cells are highly vacuolated, contain an extensive endoplasmic reticulum membrane complex, and form asexual spores by yeast-like budding.[145] Their formation can be induced by acriflavine, toluidine blue, fluorescein, basic fuchsin, ethidium bromide, and methylene blue, respectively.[153] A similar effect can be obtained by physiological manipulation of the cultivation conditions, e.g., by using media with citric acid, ammonium sulfate, and an increased content of Ca^{2+}.[154] Dickerson et al.[155] assign to the sclerotia-like cells in *C. fusiformis* the role of dominant alkaloid producers. Morphological change towards sclerotia-like growth may follow the introduction of unfavorable environmental conditions. Experiments with dyes showed that the differentiation of sclerotia-like mycelium needs not necessarily be accompanied by an enhanced alkaloid production.[153] The two processes can be separated also by mutation, low and high alkaloid producers appearing among mutants with the predominance of the sclerotia-like cells. Independence of morphology and alkaloid production was also demonstrated by Esser and Tudsynski.[62]

4. Oleosomes

An extremely large number of oleosomes is typical of the submerged cultures of *C. purpurea*. Some observations by Voříšek and Sigler[156] imply that, in moments of need, cells are capable of creating *de novo* membranous structures from nonstructured lipid pools (droplets) and protein present in the cell vacuole. The transformation of originally apparently unstructured lipid droplets into multilayered lipid-protein-lipid structures proceeds from the droplet surface which is initially bound by a ''half-unit'' (single fineline) membrane, not by the usual bileaflet membrane which could be expected to form primer for further structurization.

The whole membranous structure in oleosomes appears to serve the sole purpose of providing a spatially and chemically optimal environment for maximum functional activity of phosphoenolpyruvate carboxylase, an enzyme which is essential for lipid catabolism. The subsequent disappearance of the lipid matrix of oleosomes results in the formation of concentric assemblies of multivacuoles or segregated vesicles bound apparently by bileaflet membranes without any catalytic activity of phosphoenolpyruvate carboxylase.

Ultracytochemical localization of malonyl-CoA formation in endoplasmic reticulum and the homologous vacuolar membranes of *C. purpurea* are correlated with the condensation of lipid deposits in the cytoplasm and in vacuoles.[156] In contrast to general opinion, the fusion of the cytoplasmic oleosomes with membrane-bound vacuolar compartments is extremely rare and is spatially dissociated from malonyl-CoA synthesis. Cytochemical localization of one of the essential reactions of β-oxidation of fatty acids (EC 4.1.1.31) inside vacuolar oleosomes revealed intermediary steps of reorganization of the oleosomes into multilayered membranous aggregates which give rise to intravacuolar unit membrane vesicles or whorls. Unit membranes are assembled *de novo* from lipid inclusions in vacuoles of *C. purpurea*.

5. Proteolytic Activity

Changes in proteolytic activity and isoenzymes (indicative of change in the rates and patterns of protein hydrolysis) occur during normal differentiation and development (also under aberrant conditions) in different organisms. This situation may be extrapolated[46] to mean that changes in the rates and patterns of protein hydrolysis result in different bioactive peptides in varying amounts which will then regulate the course of development of the cell.

With increased alkaloid formation, the proteolytic activity of the *Claviceps* mycelium decreases.[91,157] Many authors[16,158,159] observed intense proteosynthesis during alkaloid formation without a subsequent increase in the overall content of cell proteins. These results indicate a substantial protein turnover.[158] Amino acids freed during protein degradation pass into the metabolic pool of the mycelium and take part in various synthetic processes. Řeháček et al.[160] demonstrated that there is a positive relationship between the synthesis of alkaloids and the activity of glutamine synthetase converting most of the free cell ammonia into nontoxic glutamine. Glutamine also takes part in the synthesis of tryptophan — the precursor of alkaloids and inducer of their formation. The increased concentration of numerous metabolites in the cell pool can be attained by increasing the medium osmolarity.[161,162]

F. Locale of Biosynthesis

The region of formation of ergot alkaloids is not generally identical with the place where alkaloids are found in increased concentrations. The members of *Ipomoea* genus form alkaloids in leaves and accumulate them in seeds, which do not produce the alkaloids. The locale of clavine alkaloid synthesis in *Claviceps* fungus is the endoplasmic reticulum which later forms vacuoles containing lipoproteins.[60] Clavines and simple derivatives of lysergic acid are found primarily in the medium. This cannot, however, be considered as an indication of active alkaloid transport. The ratio between intra- and extracellular liquid volume is 1:25, so that 95% of the alkaloids are in the medium at equilibrium between alkaloid contents in the cell and in the medium.

In contrast, peptide ergot alkaloids (ergopeptines) are mostly intracellular. Cellular compartmentalization may help to separate ergopeptines from potentially sensitive sites in the cell. Besides vacuoles, the endoplasmic reticulum, and cell walls, other compartments also exist which contain negatively charged groups able to bind ergopeptine cations. The main storage compartment for ergopeptines is lipid droplets (oleosomes) in which alkaloids are accumulated without any influence on metabolism.[163] During germination of the sclerotial cells, oleosomes are phagocytized by the developing vacuoles and the lipids are metabolized inside the vacuoles. Also, alkaloids might be metabolically degraded in vacuoles. Robbers et al.[80] feel that enzymes degrading alkaloids of ergot are activated by phosphate.

In parasitic *Claviceps* cultures, cytoplasmic oleosomes fuse[164] while in young saprophytic cultures they do not coalesce, possessing at least a superficial monomolecular phospholipid layer.[165] In contrast, vacuolar oleosomes have no visible boundary. The occurrence of enzyme activity inside the oleosomes is not surprising since they are known to contain proteins[166] and enzymes.[167] The observed frequent contact of cytoplasmic (and vacuolar) oleosomes with endoplasmic reticulum membranes synthesizing malonyl-CoA indicate that sites of fatty acids synthesis are near or in the stained strands of endoplasmic reticulum and the stained amorphous vacuolar contents.[60,168] The finding of Řeháček and Kozová[169] showed that the synthesis of malonyl-CoA was enhanced in the period of alkaloid synthesis. Neumann et al.[163] reported that oleosomes are the main storage compartment for alkaloid in *C. purpurea*. From the above facts we conclude that the staining of endoplasmic reticulum membranes might represent synthesis of malonyl-CoA for both lipid and alkaloid synthesis.

In *C. purpurea,* acetyl-CoA carboxylase activity was detected only in cells producing high amounts of clavine alkaloids.[168] It is confined to the membranes of endoplasmic reticulum and their derivatives: tonoplast of vaculoles and tiny vesicles plus amorphous lipoprotein material inside vacuoles (Figure 22*). It is suggested that the activity of acetyl-CoA carboxylase is limited to the synthesis of clavine alkaloid precursors, which occurs in the endoplasmic reticulum and its derivatives.

If ergopeptines accumulate in vesicles formed by invaginations of membrane, they could then be secreted through cell walls and never re-enter the cell. Enzymes catalyzing terminal steps of alkaloid synthesis could be associated with membranes as in many other microbial enzymes.

G. Alkaloid Role

The biosynthesis of ergot alkaloids requires specific enzymes. Because alkaloids are accumulated in significant amounts, a significant pool of materials of primary metabolites and energy is used in their formation. Therefore, alkaloid synthesis is suggested to have a function to prevent its elimination during evolutionary selection.

The biosynthesis of excessive amounts of alkaloids may be energetically favorable for the utilization of metabolites accumulated as a result of a faulty regulation of primary metabolism. Ergot alkaloids are not waste physiologically inert products of cell metabolism.[170-172] Their biosynthesis plays a role in the physiology of the production organism by maintaining certain mechanisms of metabolism and cell division in an operative state, especially under conditions of limited growth of the organism.

* Figure 22 appears after page 136.

In a productive cell, ergot alkaloids increase the activity of tryptophan synthetase and acetyl-CoA carboxylase, and inhibit anthranilate synthetase, isocitrate lyase, malate synthetase, and adenosine triphosphatase.[91,98] Suppression of the peptidase activity was also observed. Ergometrine added to a submerged culture of *Claviceps paspali* increases the alkaloid formation by 120%.[170,171] It is noteworthy that ergometrine also enhances the production of rhodomycins by *Streptomyces purpurascens*.[173] Ergotamine and ergometrine reduce the intensity of conidia formation in a production culture of *C. purpurea*.[152] The effect of tryptophan on the enzyme system in ergot alkaloid synthesis[78,81] suggests the participation of alkaloids in the feedback inhibition and repression of the enzyme-forming system. Submerged cultures of *C. paspali* can double the production of amide derivatives of lysergic acid if supplemented with elymoclavine, an early intermediate. This is not merely stimulating alkaloid synthesis; enzyme capacity rather than precursor supply is the limiting factor.

The formation of clavine alkaloids is a part of endogenous metabolism releasing the energy required for maintenance of the integrity and viability of cells under conditions of limited growth.[170,171] As a part of important regions in the productive cell this synthesis is more resistant to regulation noise. Clavine alkaloids probably form a reservoir from which, in the course of evolution, interactions develop, leading to the formation of peptide alkaloids.

Investigations of the influence of ergot alkaloids on microorganisms have been restricted to their influence on the metabolism of *Claviceps* spp., i.e., those fungi which produce ergolines themselves.[92] Investigating the transformation of ergot alkaloids by other microorganisms, Schwarz and Eich[173] observed that the addition of agroclavine to cultures of certain bacteria caused a pronounced growth-inhibiting effect. This observation was also made with some human pathogenic bacteria. The growth of *Streptococcus aureus* was completely inhibited by a concentration of 80 μg/mℓ. In the case of rhodomycins producing *Streptomyces purpurascens*, agroclavine inhibited both the culture growth and the antibiotic production. Elymoclavine inhibited the growth of *S. purpurascens* but, unlike agroclavine, it increased the production of rhodomycins.

V. REGULATION

Regulation is an essential condition of life. Hardly any reaction or pathway can be imagined to be adaptive in the absence of regulation. Complex regulatory responses result from the simultaneous occurrence of numerous regulatory interactions, each relatively simple in principle. Activities in *Claviceps* cells are associated with the action of enzymes that, in order to function, are situated in the proper cellular environment. This environment, for most enzymes, is within the cytoplasm or closely associated with the cytoplasm by attachment to membranes or in equivalent position in association with an organelle. The activity of enzymes is affected either directly or indirectly by the redox potential of the immediate cellular environment. Most enzymes can be inactivated and restricted in activity by oxidation or reduction of functionally important groups of ions in or about the active center of the enzyme. Redox-related events may also change the binding of ligands, ions, or cofactors and thus change activity in a less direct manner. It is evident that the redox state of the *Claviceps* cytoplasm is thus critical for the activity of coordinated enzymic events such as protein synthesis, cytoplasmic movement, general catabolic and anabolic metabolism, and ATP production. In addition, without changes in cellular redox levels, many more complicated activities, such as cell division, cell enlargement, and growth and differentiation may be blocked. Any protein or enzyme that does not function in harmony with the prevalent cellular redox levels will be inactive; thus, any protein synthesized *de novo* must either function under existing conditions or be able to alter cellular redox levels to provide a more suitable environment.

If a system exists in *Claviceps* cells that can, through interaction with growth regulators

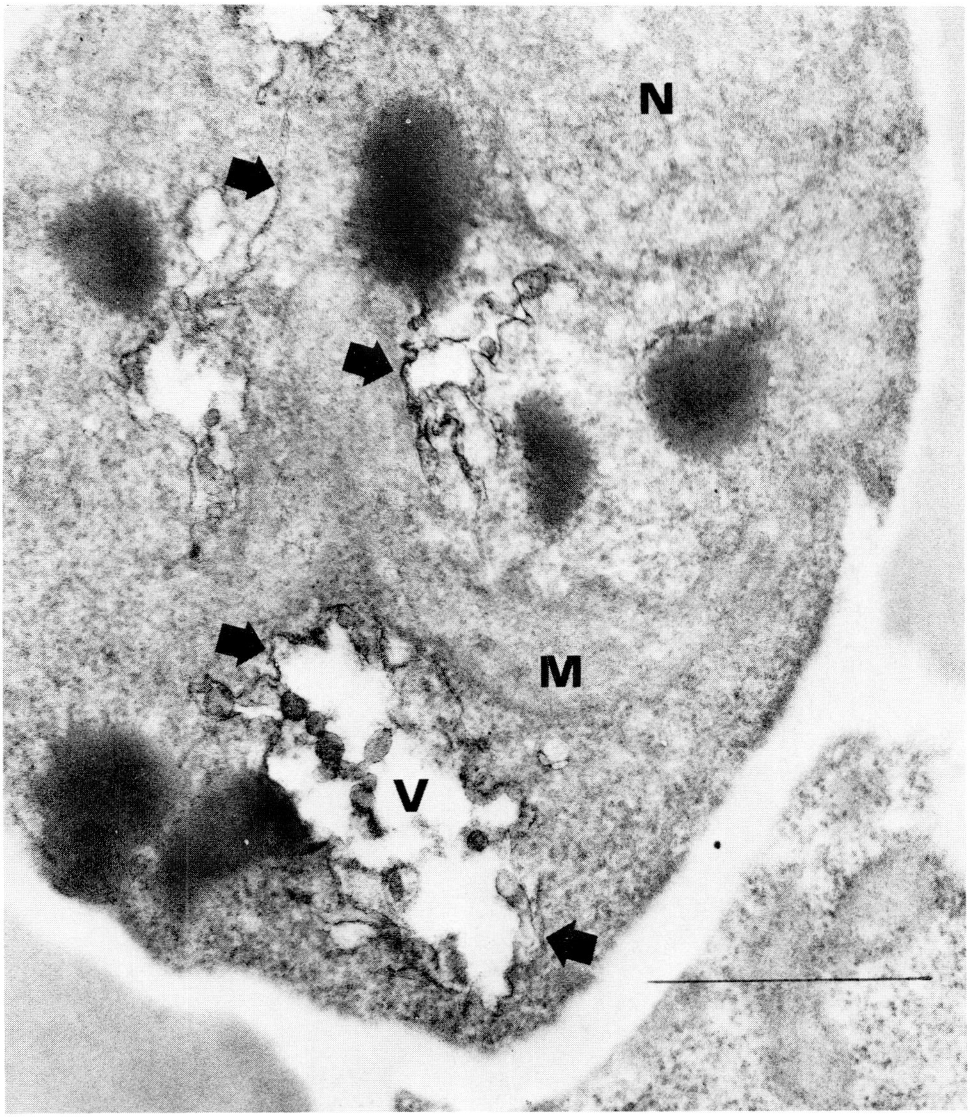

A

FIGURE 22. The ultrastructural localization of acetyl-CoA carboxylase activity in *Claviceps purpurea* producing clavine alkaloids.[168] (A) M = mitochondria; V = vacuole; N = nucleus. Arrows indicate vacuolar membrane vesicles. Magnification × 50,000. (B) ER = endoplasmic reticulum; V = vacuole; L = lipids. Magnification × 85,000.

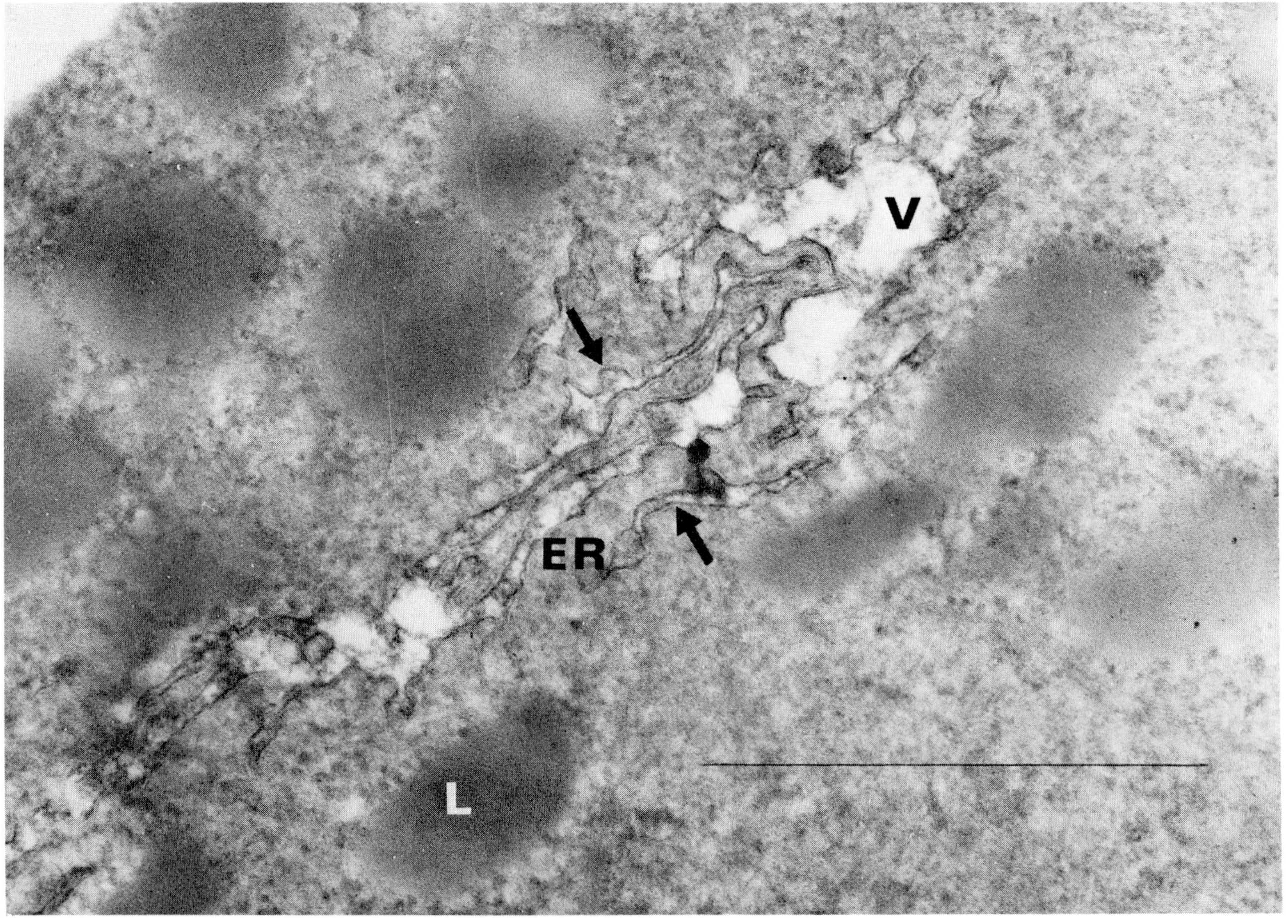

FIGURE 22B.

and environmental stimuli, regulate cytoplasmic redox levels, then this may constitute a primary mode of metabolic and developmental control. Cell activities such as *de novo* enzyme biosynthesis, cell division, and differentiation would then become secondary and tertiary effects resulting from redox level changes.

Changes in the redox potential of the cytoplasm may exert effects secondarily on the nucleoplasm; thus, the following very general changes may occur:[174] (1) the amount of the specific regions of DNA available for transcription may be altered because of direct conformational changes of protein molecules binding into the DNA; (2) activation of enzymes such as phosphorylases which may in turn control the metabolism of the molecules covering DNA; (3) activation of enzymes that actually unwind DNA, e.g., DNA seivelase and DNA gyrase.[175] Changes in redox potential may cause the migration of molecules previously in the cytoplasm to the nucleus which may, in turn, act to further regulate the expression of the uncovered regions. In this manner, previously silent regions of DNA may become activated.

The resultant RNA (and eventually proteins) formed, however, result from carefully controlled and sensitive mechanisms. It is becoming clear that the variation in the transcription of structural genes may be limited and that the three-dimensional structure of nuclear RNAs may be an important factor in determining mRNA formation. Thus, changes in nuclear redox levels that would affect these structures may be a primary mode of control. Nuclear enzymes may copy all the DNA available for transcription; the final mRNAs formed would be due to selective metabolism related to three-dimensional structures of the initial transcript.

Different levels of redox level changes are seen as a mechanism for sequential activation of gene batteries that, acting in unison, result in specific or general cellular activity that may in turn prepare the cell for further changes by a cascade of reactions and a sequential exposure of the genetic abilities of the cell. The end point of these changes may either result in the formation of a specific cell type (differentiation) or may be cyclic in nature, resulting in the return to the initial conditions (i.e., cell division). The redox potential of the cytoplasm may further determine the three-dimensional structure of such molecules as heterogenous nuclear RNA (hnRNA), mRNA, or ribosome components. The regulation of the many diverse reactions mediated by secondary messengers such as cyclic guanosine monophosphate (cGMP) or Ca^{2+} may also relate to cellular redox levels in *Claviceps*.

Ergot alkaloids are formed from intermediates or end products of primary metabolic processes. Therefore, their biosynthetic origin could be a basis for examining their regulation. Enzymic reactions involved in the genesis of ergot alkaloids compete for the same coenzymes and ions that catalyze basal metabolic reactions. Substractions of these compounds from the common pool affect other processes in which these coenzymes and ions take part. The mechanisms which determine the production of different alkaloids lie mainly in four factors:[42]

1. Genetic blocks in some reactions of the biosynthetic pathway, which determine the accumulation of different alkaloid intermediates.
2. Scarcity of one of the amino acids of the peptide sidechain, which causes a limitation of the related peptide alkaloid, resulting in the accumulation of lysergic acid and of single lysergic acid derivatives.
3. Specificity of the biosynthetic enzymic systems, which is shown by the following evidences:
   a. Only amino acids with nonpolar groups can be incorporated into peptide alkaloids, which limits the number of available amino acids to seven.
   b. The degree of enzyme-substrate affinity is required which is different in the three different possible positions of incorporation.
4. Competition among amino acids for a common limiting precursor.

According to cybernetics, ergot alkaloids are formed, like other substances, in at least two ways; otherwise, regulation of their formation would not be possible.

The organisms producing clavine alkaloids are biocybernetically similar whereas the biocybernetics of the organisms producing peptide ergot alkaloids are much more specific. This is also reflected in the fact that peptide alkaloids have so far been found only for one species, i.e., *Claviceps purpurea*. In biosynthetic schemes of ergot alkaloids, even the exact order of reactions may not be important. Several parallel pathways may convert a given intermediate to the same end product. The channeling of metabolic intermediates in the desired direction is aided by the knowledge of the regulatory mechanisms and biosynthetic pathways.

Since the maximum potential rate of energy production by *Claviceps* spp. is much greater than the requirement for ATP, then fungus has evolved mechanisms to ensure integration, correlation, and control of cellular processes. Its metabolism is regulated by mechanisms which function at various biochemical levels and time scales. These range from coarse genetic control of enzyme synthesis and activity (i.e., induction, repression, derepression, etc.) to fine metabolic turning effected through kinetic and thermodynamic aspects of the systems.

The ergot alkaloid skeleton, ergoline, is derived from tryptophan, mevalonic acid, and methionine.[176-178] Krupinski et al.[84] showed that thiotryptophan, which is inactive as a precursor of the alkaloids, was as effective as L-tryptophan in increasing alkaloid production. L-Tryptophan and thiotryptophan increase the level of DMAT synthetase, suggesting that L-tryptophan induces the production of this enzyme and possibly that of other enzymes in the pathway.

The treatment of ergot alkaloid precursors depends on the biocybernetics of the given individual, i.e., on the appropriate genes being present in the cell, on their activation and deactivation in dependence on changes in the medium, and on structural changes. End-product regulation of alkaloid synthesis involves feedback inhibition.[179] Repression of enzyme synthesis by the end product may be of lesser importance for the regulation of alkaloid formation. Besides an enzyme being inhibited or inactivated, it can be modified, altered, or degraded. Concentrations of cellular enzymes are regulated so that neither too much nor too little is produced by normal cells. It has not been proved that the "correct" amount of enzymes are made by cells engaged in alkaloid synthesis.

The growth of *Claviceps* strains is ultimately dependent upon innumerable enzyme reactions and the successful integration of many metabolic control systems. Consequently, the implementation of any unifying principle will undoubtedly yield numerous exceptions when applied to the study of *Claviceps* under a variety of growth regimes. Accelerated growth and alkaloid synthesis are obtained with surface-active agents which increase nutrient uptake by altering membrane permeability.[180] They affect the timing but not the maximum yield of products.

Production of ergot alkaloids is inhibited when glucose is used as a carbon source. However, this does not show that the inhibitory effect of glucose on alkaloid synthesis is attributable to catabolite repression. Results of kinetic measurements of phosphofructokinase (PFK) in desalinated cell-free extract from an axenic *C. purpurea* culture producing clavine alkaloids[103] indicate that the inhibition of PFK by glucose or some of its related metabolites points to a feed-forward regulation of PFK (Figure 19) and to a possible new regulation model of the glycolytic pathway. PFK exhibited a hyperbolic kinetics which was not typical of this enzyme.

Industrial microbiologists have solved the problem of catabolite repression by medium manipulation. One way is to use two carbon sources in the fermentation. A low concentration of a readily utilizable carbon source allows good growth of the organism. A second non-repressive slowly utilizable carbon source is metabolized during production of the secondary metabolite. Another method is the continuous feeding of the repressive carbon sources at such a slow rate that the inhibitory effect is not exerted.[181]

Knowledge of the regulation of amino acid biosynthesis in alkaloid producers is crucial for the handling of industrial strains. Amino acids are needed for protein synthesis and culture growth and are important precursors of ergot alkaloids. The flow of intermediates into the metabolic pathway leading to peptide alkaloids can be modified either through genetic blocks in some unidentified step of the pathway itself or by the shortage of one of the amino acids of the peptide sidechain of the alkaloids produced.[20] The flux through single amino acid pathways or a group of interconnected branched pathways is regulated primarily in a system of flood gates through feedback inhibition at the entry into these pathways.[182] Regulation of enzyme synthesis is usually controlled through the systems of "general control of amino acid biosynthesis" and serves the cell to cope with imbalance of the external amino acid supply. The combination of regulatory mutations affecting either feedback control of the activity of entry enzymes or the "general control" can be used to enhance the flux through a desired pathway considerably. Further increases in enzyme levels and possibly in the flux can be achieved through gene cloning.

During *Claviceps* spp. growth, regulatory controls operate at the level of transcription of DNA into RNA, the translation of RNA into polypeptides, and the activity of enzymes. However, in alkaloid biosynthesis, additional specific control signals are superimposed on those regulating microbial growth and general metabolism. Practically no experiments have been performed that can identify the type of transcriptional and translational apparatus that exists in cells actively engaged in alkaloid formation. Two types of transcriptional apparatus may exist simultaneously. One would transcribe log phase genes and the other would be capable of transcribing genes for alkaloid synthesis. The need for inducer early in the growth phase, well before the enzymes are made, is unexpected and suggests that the appropriate genes are transcribed during the exponential phase of the culture growth. Their subsequent expression may be due to translational control. Protein turnover in late growth-phase cultures undoubtedly indicates the synthesis of enzymes catalyzing alkaloid synthesis as well as extensive differentiation of cellular structures then going on.

Mechanisms that control inactivation of alkaloids in the producing organism have commercial significance because their manipulation can be used to increase the yield of alkaloids in industrial fermentations. Enzymes that convert alkaloids into biologically inactive compounds influence the final yield of alkaloids accumulated in the fermentation broth.

Some genetic block might be the basis of the chemical differentiation between *C. purpurea,* which produces peptide alkaloids, and *C. paspali,* which produces lysergic acids or its simple derivatives. The ratio amount of the various groups of alkaloids formed can also be affected by the availability of some amino acid of the alkaloid peptide side chain, which in turn might be controlled both by genetic and environmental factors.

Enzymes from several different amino acid biosynthetic pathways are under a common control and derepress when any amino acid becomes limiting.[183] For example, starvation for histidine leads to derepression of enzymes in the histidine, arginine, tryptophan, and lysine pathways. This global regulatory response, known as general amino acid control, affects at least 24 different amino acid biosynthetic enzymes in six different pathways. Analysis of a set of *trans*-acting regulatory genes indicates that genes subjected to general control are regulated by both positive and negative regulatory elements.

To understand selective gene expression during cell development, we must understand how the inducers and repressors which orchestrate genetic activity are distributed within a cell. We must also understand the cellular and intercellular distribution of gene products with respect to the distribution of the substrates, effectors, and other cellular components upon which gene expression depends. The protoplast technique and especially the methods of gene transfer via protoplast fusion have opened up new possibilities in both the cellular and molecular aspects of fungal genetic engineering.

Analysis and evaluation of corresponding experimental data from several laboratories have revealed that mitochondrial (mt) DNA, regardless of whether it originates from mt plasmids or from the mt chromosome, may be used to construct autonomously replicating vectors for molecular cloning in eukaryotes. Vectors of this kind replicate efficiently, are maintained stably in the host, and therewith fulfill the prerequisites for practical application.[184]

VI. BIOTRANSFORMATION

Microbial systems are also useful in carrying out synthetic reactions and show certain advantages over classical synthetic reagents when applicable. First, reactions occur at chemically "nonactivated" positions. One often observes regional selectivity with polyfunctional molecules, a major advantage since many chemical reagents cannot distinguish multiple similar functional groups. A high degree of stereoselectivity (substrate and product, or both) is often observed. A final advantage, and one most important to the chemist, is the predictability of these reactions. This area is rapidly moving toward a more refined approach to choosing organisms for their predictable type-reaction specificity.

Much of the work on alkaloid transformations was motivated by the desire to produce more effective drugs from naturally available compounds. No doubt it was encouraged by the success of similar efforts in the steroid field. Of the numerous transformations of ergot alkaloids now demonstrated, a high proportion is brought about by *Claviceps* cultures. They represent steps in the complex branched biosynthesis process by which the alkaloids are formed. Certain members of the clavine series, notable chanoclavine-I, agroclavine, elymoclavine, and probably also paspalic acid, are mainline intermediates en route to substituted lysergic acid. Conversions along this pathway in a forward direction have been shown to occur in alkaloid-producing cultures,[185] but not with other microorganisms. The enzymes used in these conversions have narrow specificity and a restricted distribution.

Chanoclavine markedly increased the production of the pathway end products, ergometrine and ergotamine.[186] Submerged cultures of *Claviceps paspali* can also double production of amide derivatives of lysergic acid if supplemented with elymoclavine, another early intermediate.[187] Elymoclavine disappears from the culture as the product forms so that it is not merely stimulating alkaloid synthesis.

The variety of clavine alkaloids which have been formed in *Claviceps* spp. and other fungi-producing ergot alkaloids is, to a great degree, due to the action of enzymes that direct elymoclavine and agroclavine from the main biosynthetic route. Such shunt activity is particularly evident in strains that lack the complete pathway and are unable to synthesize lysergic acid amide.[188] Accumulated $\Delta^{8,9}$ intermediates may be reduced to dihydroalkaloids or converted by allylic hydroxylation into a variety of products.[189] The oxidative conversions are catalyzed by peroxidase.[190] Elymoclavine first forms a 4:1 mixture of 10 α- and 10 β-hydroxyelymoclavines. Under prevailing acidic culture conditions, these rearrange to give peniclavine and isopenniclavine. A similar series of events may account for the conversion of agroclavine into setoclavine and isosetoclavine. Horseradish peroxidase has been shown to catalyze the conversion in vitro.[191] Thus, it is no surprise to find reports of the occurrence of these conversions in plant homogenates as well as in other microbial cultures (Table 7). In the psilocybin-producing fungus *Psilocybe semperviva,* conversion of $\Delta^{8,9}$ into 8-hydroxy-$\Delta^{9,10}$ alkaloids is about 80% and some stereoselectivity is observed.[194] Although elymoclavine gives a 3:2 mixture of isopenniclavine and penniclavine, agroclavine is exclusively converted into setoclavine. The fungus rapidly oxidizes the lysergic acid derivatives ergometrine and ergotamine in the indole ring. No identifiable products have been recovered, but Yamatodani et al.[192] were able to isolate 2-hydroxyagroclavine from cultures of *Corticium sasakii* to which agroclavine was added.

In other fungi and bacteria, several transformations unrelated to the dominant biosynthetic conversions of producing cultures have been observed. Strains of *Aspergillus fumigatus, Corticium praticola,* and *C. sasakii* are reported to reduce elymoclavine to agroclavine. *A.*

Table 7
CONVERSION OF AGROCLAVINE (A) AND ELYMOCLAVINE (E) BY MICROORGANISMS BESIDES ERGOT FUNGI[192,193]

Microorganism	Substrate	Conversion product
Streptomyces reseochromogenes		Setoclavine
S. rimosus	A	Isosetoclavine
S. purpurescens		
S. vinaceus		
S. lipanii	E	Penniclavine
S. flaveolus		Isopenniclavine
Fusarium lini	A	Setoclavine
F. oxysporum		Isosetoclavine
F. niveum	E	Penniclavine
		Isopenniclavine
Corticium sasakii	A	Setoclavine
		Isosetoclavine
		2-Hydroxyagroclavine
	E	Penniclavine
		Isopenniclavine
Penicillium roqueforti	A	E
P. concavorugulosum		Chanoclavine
	E	A
		Chanoclavine
Aspergillus conicus	A	Chanoclavine
	E	A
A. fumigatus	A	Chanoclavine
	E	A
		Chanoclavine
		Lysergol

fumigatus also forms lysergol, lysergene, and chanoclavine in a series of reversible reactions[195] but, since this fungus is known to be capable of producing ergot alkaloids, some of these conversions may not be unrelated to its biosynthetic potential. Lysergol is slowly converted by allylic oxygenation in cultures of *Acetobacter aceti* into penniclavine along with traces of isopenniclavine, alymoclavine, and an unidentified product. *Pseudomonas aeruginosa* metabolizes agroclavine to penniclavine, presumably by two allylic oxygenations.

Radiotracer studies with *Claviceps* cultures have indicated that *N*-methylation and *N*-demethylation reactions are common.[196] Removal of methyl groups may, at least in part, be due to the activity of peroxidases. Noragroclavine is formed from agroclavine by several *Streptomyces* species[192] and accumulation of norsetoclavine[197] and *N*-demethyl-chanoclavine-II[198] occurs under conditions where allylic hydroxylation is also prominent.[198]

Some of the more interesting microbial conversions of alkaloids have been carried out with semisynthetic drug. A subsidiary, if not prime, motive is often the prospect of specifically modifying the drug to improve its activity or decrease undesirable side effects.

VII. STRAINS

In some cases the saprophytic production of ergot alkaloids is more economical than the parasitic cultivation on rye plants. Alkaloid yields in bioreactors, as a percentage of the mycelial dry weight, may exceed 20%, whereas natural sclerotia contain less than 1%, or even less than 0.1%. In bioreactors the process is more precisely controlled, the cultivation is not limited to the specific season, and the production and isolation of alkaloids can take place on the same site. Furthermore, it is possible to apply some manipulations which are

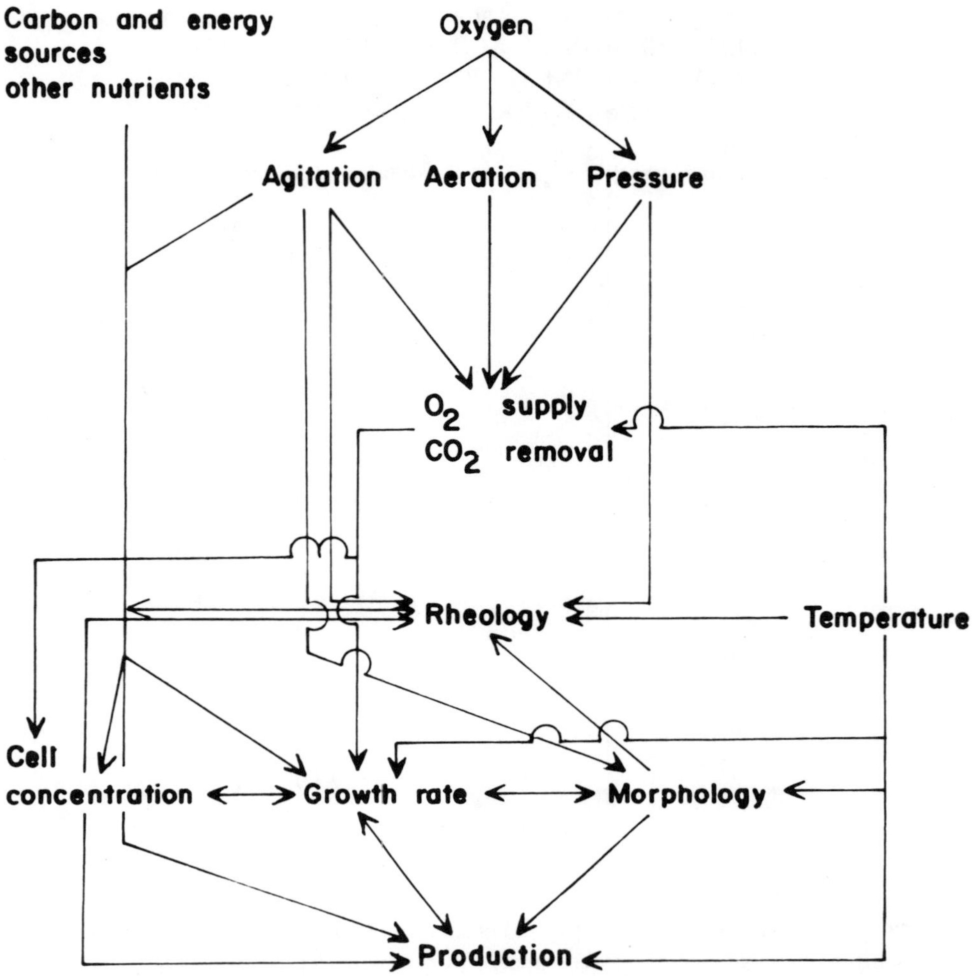

FIGURE 23. Possible manipulation of cultivation conditions, culture growth, and development, and alkaloid production in a bioreactor.

not practicable under field conditions (Figure 23). Clavines and simple lysergic acid derivatives are easier to manufacture than ergopeptines.

A. Scale-Up

In the course of optimization and scale-up of the process to production levels, certain facts must be taken into account and several difficulties of biological and technological nature must be overcome:[39]

1. The bulk of the peptide alkaloids is located in the mycelial cells and is not excreted into the medium. No satisfactory method to achieve this excretion has yet been found. Conditions for a good yield are therefore large mycelial mass per liter with a high alkaloid content.
2. The long cultivation period of 2 to 3 weeks and the absence of any antibiotic formation in *Claviceps* cultures require very high standards of sterility in operation and equipment.
3. The sensitivity of the organism against stress and the necessity for high oxygen tension require a well-balanced system of aeration and stirring.

4. The fermentation reacts very strongly to small alterations in the media composition, especially when antifoam agents must be added. Most antifoams cause considerable loss in alkaloid yield.
5. High alkaloid-yielding mutants of *Claviceps* are generally difficult to preserve and sometimes these same mutants will lose their high production activity. It is therefore necessary to keep the production strains under constant observation.

The saprophytic alkaloid synthesizing strains are the prerequisite for modern production of ergot alkaloids. Starting from the alkaloid content, some groups of strains are worth mentioning:[199]

- Clavines *C. fusiformis* B35
  C. purpurea Pepty 695/ch
  C. purpurea 129
  C. paspali SD 58
- Simple lysergic acid derivatives *C. paspali* NRRL 3027
  C. paspali Li 342/SE60
- Ergopeptines
  Ergotamine *C. purpurea* ATCC 15383
  C. purpurea FE 275
  C. purpurea OKI 56/1970
  Ergosine *C. purpurea* MUT 168
  Ergocristine *C. purpurea* ATCC 20 103 (IMI 131669, FIS 40)
  C. purpurea Ech K 420
  Ergocornine and ergosine *C. purpurea* FI 43/14
  Ergotamine and ergokryptine *C. purpurea* FI 32/17
  C. purpurea IC/39/20
  Ergocornine and ergokryptine *C. purpurea* Exy 20
  C. purpurea OKI 00022
  C. purpurea D-M838 (IMETPA 130)
  C. purpurea Pepty 695
  Dihydroergosine *Sphacelia sorghi*

B. Degeneration

Alkaloid production is frequently lost. The polygenic character of alkaloid biosynthesis may be partly responsible for the frequency of genetic degeneration of industrial strains. The term ''degeneration'' is applied to the phenomenon whereby a strain decreases in productivity of alkaloids after repeated transfer on agar media. The morphological development of the mycelium in submerged *Claviceps* cultures shows significant differences between high-yielding and degenerated cultures. Thick walled non-sporulating round cells with a full cytoplasm develop in producing cultures, while in nonproducing cultures the mycelia are composed mainly of long, thin, vacuolated cells which sporulate abundantly. Degenerated strains are no longer able to use high amounts of nutrients while the alkaloid yield of more productive mutants could be further improved by supplying them with higher carbon and nitrogen concentration.

The fact that *Claviceps* cells are polynuclear also contributes to the explanation of the phenomenon of strain degeneration. Actively growing hyphae of *Claviceps* spp. have at least 10 nuclei per separated cell and as many as 46 have been recorded in the growing tip.[81] In *C. fusiformis,* the number of nuclei per cell decreases as the culture ages, and even hyphal tips become mononucleate, yet this strain produces a high yield of alkaloids during late growth.[81] During cell growth the individual nuclei may divide at various rates. A shift in the genetic configuration thus occurs, leading to a shift in the phenotypical manifestation of the population. Changes in the strain ability to produce alkaloids occur not only due to genetic changes, but also due to exceptional sensitivity of the strain to changes in cultivation medium.

A characteristic feature of alkaloid-producing strains of both *C. paspali* and *C. purpurea* is their instability, whereby they revert to nonproducing forms after only a few subcultures. In contrast, the clavine alkaloid-producing strains of *C. fusiformis* are usually more stable in submerged culture.[200] Alkaloid synthesis by *C. fusiformis* appears not as closely related to sclerotial cells as had been demonstrated in *C. purpurea*.[143] The production activity of the saprophytic cultures of *C. paspali* FA did not substantially change on storage for 8 weeks on Sabouraud's solid medium at 4°C. However, it decreased by 50% after 15 weeks. In a long maintained culture, (4°C) recovery of growth and metabolic activity was reached after a two- to threefold transfer of the culture at short 3-week intervals.[77]

C. Improvement

The production of ergot alkaloids is a result of dynamic conversion taking place during the development of the culture. Inside the cell the process of development is determined by interaction and/or competition of two interconnected dynamic systems.[201] The first system is fast and is represented by enzymically catalyzed reactions of small molecules.[202] This system can be imagined as a motor of culture evolution. The second dynamic system is joined with cell structures determined by macromolecules as nucleic acids, proteins, etc. This system is relatively slow and acts as a driving filter. The resulting behavior of the culture or the permanent transformation of metabolic field is then determined by the interaction of both these dynamic systems in dependence on the environmental conditions, i.e., on the total mass and energy balance of the culture. In order to increase alkaloid production, it is possible to apply two methods. The first method modifies the structure of the driving filter by means of genetic manipulations. In this way the slow dynamic system is affected. The second method takes account of an organism having the structure of its slow dynamic system already stable. In this case the form of the metabolic field is altered by means of physiological and technological manipulation, e.g., by addition of a nutrient or another substance, by temperature, aeration, pH control, etc.

Mutagenesis followed by subsequent selection of a strain producing superior yields is an important technique in increasing the fermentative yield of alkaloids. One of the first reports of Kobel et al.[203] describes a mutant strain of *C. purpurea* which produces an increased amount of alkaloids. Subsequently, Kobel and Sanglier[39,204] obtained an ergocristine-producing *C. purpurea* strain which showed a 180-fold increase in alkaloid production over the parent strain after an eight step mutation-selection with different mutagens. Their selection technique involves the isolation of 1000 colonies after each mutation and the subsequent testing of each isolate in shake cultures for alkaloid production. It is thus a time-consuming process.

Screening for *Claviceps* mutants is difficult. Ergot alkaloids are not essential to the life cycle of the producing organisms; hence, conditional lethal protocols cannot be used. Although protoplast fusions have been used to create recombinant strains in the past, most industrial strains with improved yields have been obtained by a combination of brute force mutagenesis and selection. Some of the known chemical mutagens applied are: *N*-nitro-*N*-nitroso-*N*-methylguanidine (NNMG), ethylmethansulfonate (EMS), methylmethansulfonate (MMS), ethyleneimine, nitrous acid, etc.

The difficulty of producing stable mutants from ergot strains that have multinucleated cells and that do not readily produce conidia in culture, such as an ergotamine-producing strain, was overcome by first forming protoplasts of the fungus and then subjecting them to the mutagen.[205] Stable auxotrophs were obtained in this manner.

Among various strains of *C. paspali* tested, those that were good alkaloid producers were the best candidates for increasing yield.[188] Since elymoclavine does not boost the yield from low producers to that of high-producing strains, enzyme capacity rather than precursor supply must be the limiting factors in the former.

Auxotrophic mutation generally leads to lower productivity of organisms forming alkaloids. All auxotrophic mutants of *Claviceps* strain SD 58 that produced high alkaloid levels after a second mutation reverted to prototrophs.[205] Reversion resulted in an alkaloid production level in the mylelium that was $\leq$ 1.6 times that of the original prototroph. Mutants which retained their auxotrophic character were poor alkaloid producers. A possible explanation for the increase in the production of alkaloids by auxotrophic revertants is that the revertants are feedback-resistant mutants in which the production of primary metabolite precursor is no longer regulated, resulting in its increase. If the amount of this precursor is rate limiting in biosynthesis, there would be an increased production of alkaloids.

Rational selection procedures based on known or probable biochemical mechanisms are successfully used for many established fermentation organisms prior to fermentation screening and should now largely replace large-scale empirical mutation and selection procedures. The following selection techniques can be considered:

1. Direct colony selection
2. Selection for mutants resistant to toxic precursors or to toxic levels of end product
3. Use of analog-resistant mutants which overproduce rate-limiting biosynthetic intermediates
4. Selection of auxotrophic strains deficient for biosynthetic intermediates followed by prototrophic reversion
5. Selection for recombinant strains following protoplast fusion
6. Selection of strains which are insensitive to carbon catabolite repression
7. Recombinant DNA technology offers future prospects for the systematic construction of strains capable of overproducing existing or structurally modified alkaloids.

Targeted selection of *Claviceps* recombinant colonies through use of gene dosage effects might be considered. Dosage compensation is a relatively uncommon aspect of gene regulation. Generally, increasing the copy number of an individual gene (or genes) results in a corresponding increase in the products of that gene.

Studies of selection in chemostat, i.e., in the particular continuous-culture device, have revealed a number of important phenomena. The mutants selected in replicate chemostats are highly predictable and have a defined genetic and physiological basis.[206] This opens the prospect of controlling or perhaps eliminating undesirable genetic changes that commercially important strains sometimes undergo in prolonged continuous culture. The behavior of cloning vectors in chemostats is a subject of great interest because of the potential uses of continuous cultures for the study of the functional effects of defined, engineered mutations and for potential applications in industrial microbiology.

Differences in yield and variations within a qualitatively characteristic alkaloid spectrum testify to the importance of environment. Alkaloid production is maintained in vivo through an association with virulence. Passage of both *C. paspali* and the host selects for high-producing strains. Although the association may not be mandatory, and the spread of alkaloid production in re-isolated cultures can extend below as well as above the level of the starting culture, it is an effective vehicle for strain improvement.

In order to improve the level and spectrum of alkaloid production, a chromosomal genetic analysis of the fungus was established. It was found that despite self-fertility of the fungus, several crosses may be conducted for recombination of genetic material.[62]

At present there is a trend in biotechnology to produce ergot alkaloids through axenic cultures in bioreactors rather than through parasitic cultures in the field. The domesticated strains regularly lose sexual and asexual (formation of conidia) fertility.[207] Therefore, an improvement of these infertile strains is not possible, either by sexual or parasexual chromosomal genetics. Accordingly, Tudzynski et al.[208] attempted to develop an extrachromosomal system for genetic manipulation of such sterile strains, since in related filamentous fungi, i.e., *Podospora anserina* and *Cephalosporium acremonium,* it was shown that mitochondrial (mt) DNA or mt plasmids may serve as the basis for development of a eukaryotic cloning system.[209] A further reason to follow this line of investigation comes from the fact that mitochondrial metabolism is involved in alkaloid synthesis[172] as derived from studies of catabolite repression and of mitochondrial inhibitors.[210]

Tudzynski et al.[208] discovered plasmids associated with the mitochondria of a wild strain of *C. purpurea*. These genetic units were not formed in strains maintained in the laboratory. The mt plasmids were linear and showed homology to each other but not to the mitochondrial "chromosome". Despite localization in the mitochondria, they were not an integral part of the mt "chromosome". The possibility of using the mtDNA and mt plasmids to establish a vector system for molecular cloning in imperfect strains of *C. purpurea* has been pointed out. The genetic code has generally been considered to be universal.[211,212] However, the genetic codes of mt DNAs from various organisms have been found to differ from those of all other organisms and even from each other.

VIII. CONCLUSIONS

The biosynthesis of ergot alkaloids is a highly ordered and coordinated event. Negative and positive regulatory controls, which are operative during balanced growth, have to be lifted with the simultaneous accumulation and availability of initiating precursors. Ergot alkaloid formation lies at the level of genetic information apparatus and continues at the level of physiological realization. Genes involved in alkaloid synthesis are under rigorous metabolic control. Fungi and higher plants form ergolenes from the same precursors. In this connection it is important to determine whether the ergolene pathway was formed twice in nature independently or only once, the necessary genetic information being transferred from the fungus to the plant or vice versa. The concept of universal existence and limited expression of this genetic information is less probable.

Formation of ergot alkaloids is to a greater degree a result of change in the physiological state than an indication of a new physiological state.[170] Culture growth and alkaloid synthesis should be understood as processes which compete for key metabolic intermediates rather than as mutually exclusive phenomena. Ergot alkaloids are not waste, physiologically inert products of cell metabolism.[91,170,172] Mathematical modeling extended by physiological aspects could open a new area for a more exact and utilizable mathematical description of phenomena observed during culture growth and alkaloid production.

The study of slow dynamic changes[201] in a system bound with the genetic apparatus of microorganisms grown in long-term cultivation is very useful. The better knowledge and control of these processes will make it possible to direct the culture development towards such a metabolic field which is characterized by higher concentrations of alkaloids. Tightly regulated microorganisms with balanced metabolic circuits are of little use to industrial microbiologists because they do not overproduce cellular metabolites.

Important aspects of the biosynthesis of ergot alkaloids which still have to be worked out are the mechanism and enzymology of C-ring formation and the enzymology of formation of D-ring and of the peptide moiety. A cell-free system for the biosynthesis of the ergopeptines would greatly facilitate these studies. Ergopeptines are synthesized by multifunctional enzymes utilizing the protein thiotemplate mechanism. Such multienzyme complexes have high reaction efficiencies as a result of cellular compartmentation of intermediates. The amino acid sequence in the peptide moiety is dictated by the specificity inherent in multienzyme complexes.

Principal findings about the physiology of alkaloid formation will be provided by the results of the study of membrane processes in the productive organisms. Valuable data are expected from enzyme cytology, especially on functional properties and spacial organization of its components. Vacuoles are also worth studying. The first goal should be identification of the mechanism by which cytoplasm components are incorporated into vacuole. Recently, a detailed examination of lipid deposit degradation in *C. purpurea*[156] has provided new data on the conversion of cell lipid reserves into intravacuolar membrane complex.

Another area of interest for future studies involves the determination of intracellular locus of ergopeptine synthesis and the problem of distribution of alkaloids and their transport at the molecular level. Recent advances in molecular biology have generated a wave of excitement about the prospective application of novel microbiological techniques. Some other interesting problems which should be examined are

1. Shortening of the initial nonproductive phase of alkaloid fermentation
2. Immobilization of cells from the alkaloid-producing phase on a suitable carrier
3. Preparation of alkaloids by cofermentation[213]

Mutants of *C. purpurea* may produce much broader spectrum of the compounds or larger amounts of some alkaloid of lower complexity. The information available suggests that it should be possible to understand the genetic control of alkaloid production and to find the genes responsible for specific alkaloids. Two different mechanisms of gene expression appear to be involved in alkaloid formation. A short-term mechanism involves the timing at which differentiation is initiated and determination of particular pathways which are active at a given time. This is dictated by a concentration of nutrients which in turn influences the intracellular concentrations of substrates and low molecular weight regulatory effectors.[214] The second mechanism is a long-term effect dependent on changes in the enzyme composition of the cells which are under genetic control.

Any improvement of yield of alkaloids would need concerted breeding of the strains concerned. This requires as a prerequisite a knowledge of the life cycle and the ability to manipulate it in order to get controlled genetic recombination and the possibility to combine appropriate genes obtained either by mutation or by selection from natural sources. The knowledge of the regulatory biology responsible for alkaloid biosynthesis would be of help in designing rational strain improvement programs. Here, feed-forward regulation of phosphofructokinase of *C. purpurea* by glucose and its related metabolites[105] is worth mentioning as a new regulation model of the glycolytic pathway.

Ergot alkaloids, even the ergopeptines, are not direct gene products, but are synthesized by the enzymic transformation of small precursor molecules. Numerous enzymes are involved. The relevant structural and regulatory genes have not been mapped. In light of these problems, the isolation and cloning of the entire pathway of structural genes for alkaloids will be difficult. To help bridge the gap between academic and industrial genetics, further efforts should be concentrated on the following topics: (1) the construction of a plasmid that replicates autonomously in *Claviceps* spp. and which carries easily selectable markers; (2) the development of transformation systems with drug resistance markers for use in species lacking available auxotrophic mutants or in which such mutants are difficult to obtain; (3) emphasis on studying the genes controlling the regulation of entire pathways.

More thought should be directed to biochemical pathways. We would like to make two points about the study of enzymes. The first is that modern recombinant DNA techniques that allow the synthesis and insertion of defined informational sequences should permit conclusive identification of the role played by each important amino acid in a polypeptide. This can be facilitated by comparative analyses of the same enzyme in different organisms as well as related enzymes in the same organism. The second point is that enzymes, like organisms, work in communities. Interactions between enzymes in loose association or simple proximity is a rather neglected area of enzymology.

The search for new natural alkaloids involves the testing of nontraditional organisms. The more related the organisms, the greater the probability of their forming the same alkaloids. Mutation of one gene can also lead to the formation of new compounds and to structural modifications inside the cell. For example, cell components that were segregated in the parent cell can interact in the mutant.

The ready availability of lysergic acid led to a consideration in the first instance of the chemical modification of this moiety. This approach was also based on the hypothesis, which proved largely correct, that the peptide moiety of ergopeptines was not essential for the typical pharmacological activity of these substances. The series of ergot alkaloids can be widened by their semisynthetic analogs. In the preparation of practically important nicergoline,[13] attention was paid to the modification of the ergoline rest. It can be stated that in the foreseeable future new ergolene compounds will be used in medicine in addition to old ergot-derived drugs. Therefore, interest in ergot alkaloid research should remain active for years to come.

REFERENCES

1. **Berde, B.,** Ergot compounds: a synopsis, in *Ergot Compounds and Brain Function: Neuroendocrine and Neuropsychiatric Aspects,* Goldstein, M. et al., Eds., Raven Press, New York, 1980, 3.
2. **Lieberman, A. N., Kupersmith, M., Gopinathan, G., Estey, E., Goodgold, A., and Goldstein, M.,** Bromokryptine in Parkinson disease, *Neurology,* 29, 363, 1979.
3. **Lieberman, A. N., Miyamoto, T., and Battista, A. F.,** Studies on the anti-parkinsonian efficacy of lergotrile, *Neurology,* 25, 459, 1975.
4. **Martin, G. E., Williams, M., Clineschmidt, B. V., Yarbrough, G. G., Jones, J. H., and Haubrich, D. R.,** Potent dopamine agonist activity of a novel ergoline, 6-ethyl-9-oxaergoline (EOE), *Life Sci.,* 30, 1847, 1982.
5. **Liang, H. J. and Anderson, J. A.,** *N*-Acetyl-L-tryptophan in *Claviceps purpurea* PRL 1980, *Phytochemistry,* 17, 597, 1978.
6. El-Refai, Alkaloids of fungi. I. Formation of ergoline alkaloids by representative fungi, *Jpn. J. Microbiol.,* 14, 91, 1970.
7. **Agurell, S.,** Biosynthetic studies on ergot alkaloids and related indoles, *Acta Pharm. Suecica,* 3, 71, 1966.
8. **Vining, L. C., McInnes, A. G., Smith, D. G., Wright, J. L. C., and Taber, W. A.,** Dimeric clavine alkaloids produced by *Penicillium citreo*-viride, *FEMS Symp. 13,* Academic Press, London, 1982, 243.
9. **Abe, M.,** Ergot fungus. IX. Separation of an active substance and its properties, *J. Agric. Chem. Soc. Jpn.,* 22, 2, 1948.
10. **Rebek, J., Jr. and Tai, D. F.,** A new synthesis of lysergic acid, *Tetrahedron Lett.,* 24, 859, 1983.
11. **Bové, F. J., Ed.,** *The Story of Ergot,* S. Karger, Basel, 1970.
12. **Bernardi, L.,** Von den Mutterkornalkaloiden zum Nicergolin, *Arzneim. Forsch.,* 29, 1204, 1979.
13. **Hofmann, A.,** *Die Mutterkornalkaloide,* F. Enke Verlag, Stuttgart, 1964, 218.
14. **Gröger, D.,** Ergot alkaloids-recent advances in chemistry and biochemistry, in *Antibiotics and Other Secondary Metabolites,* Hütter, R., Leisinger, T., Nüesch, J., Wehrli, W., Eds., Academic Press, London, 1978, 201.
15. **Berde, B. and Schild, H. O.,** *Ergot Alkaloids and Related Compounds,* Springer-Verlag, Berlin, 1978.
16. **Bu'Lock, J. D. and Barr, J. G.,** A regulation mechanism linking tryptophan uptake and synthesis with ergot alkaloid synthesis in *Claviceps, Lloydia,* 31, 342, 1968.
17. **Pöhm, M., Kolassa, H., and Jentsch, K.,** Configuration of proline in ergotamine, *Monatsheften für Chemie,* 108, 393, 1977.
18. **Wrinch, D.,** *Chemical Aspects of Polypeptide Chain Structures and the Cyclol Theory,* Munksgaard, Copenhagen, 1965.
19. **Hofmann, A., Frey, A. J., and Ott, H.,** Die Totalsynthese des Ergotamins, *Experientia,* 17, 206, 1961.
20. **Bianchi, M. I., Crespi-Perellino, N., Gioia, B., Minghetti, A.,** Production by *Claviceps purpurea* of two new peptide ergot alkaloids belonging to a new series containing α-aminobutyric acid, *J. Nat. Prod.,* 45, 191, 1982.

21. **Weber, H. P.,** The molecular architecture of ergopeptines: a basis for biological inter-actions, in *Ergot Compounds and Brain Functions: Neuroendocrine and Neuropsychiatric Aspects,* Goldstein, M. et al., Eds., Raven Press, New York, 1980, 25.
22. **Wurst, M., Flieger, M., and Řeháček, Z.,** Analysis of ergot alkaloids by high-performance liquid chromatography. I. Clavines and simple derivatives of lysergic acid, *J. Chromatogr.,* 150, 477, 1978.
23. **Wurst, M., Flieger, M., and Řeháček, Z.,** Analysis of a mixture of ergot alkaloids by high-performance liquid chromatography. II. Cyclol alkaloids, *ergopeptines, J. Chromatogr.,* 174, 401, 1979.
24. **Flieger, M., Sedmera, P., Vokoun, J., Řičicová, A., and Řeháček, Z.,** Separation of four isomers of lysergic acid α-hydroxyethylamide by liquid chromatography and their spectroscopic identification, *J. Chromatogr.,* 236, 453, 1982.
25. **Flieger, M., Stuchlík, J., Sedmera, P., Vokoun, J., Černý, A., Řeháček, Z.,** Liquid chromatography of semisynthetic ergot preparations. I. Nicergoline, *J. Chromatogr.,* in press.
26. **Vokoun, J., Sajdl, P., and Řeháček, Z.,** Mass spectrometry determination of some natural clavines and lysergic acid derivatives, *Zbl. Bakt. II.* 129, 499, 1974.
27. **Davis, P. J., Glade, J. C., and Clark, A. M.,** *N*-Demethylation of lergotrile by *Streptomyces platensis, Appl. Environ. Microbiol.,* 38, 891, 1979.
28. **Smith, D. G.,** Unpublished data, 1983.
29. **Jones, J. B. and Beck, J. F.,** *Applications of Biochemical Systems in Organic Chemistry,* Jones, J. B., Sih, C. J., and Perlman, D., Eds., Wiley-Interscience, New York, 1976, chap. 4.
30. **Beliveau, J. and Ramstad, E.,** 8-Hydroxylation of agroclavine and elymoclavine by fungi, *Lloydia,* 29, 234, 1966.
31. **Kim, I. S. and Anderson, J. A.,** Personal communication, 1982.
32. **Maier, W., Erge, D., Schumann, B., and Gröger, D.,** Incorporation of L(U-^{14}C)leucine into ergosine by cell-free extracts of *Claviceps purpurea* (Fr.) Tul., *Biochem. Biophys. Res. Commun.,* 99, 155, 1981.
33. **Anderson, J. A.,** Personal communication, 1982.
34. **Kleinkauf, H.,** A survey of enzymatic biosynthesis of peptide antibiotics, *Int. Conf. Trends Antibiot. Res.,* (Abstr.), 1982, 22.
35. **Kurakashi, H.,** Biosynthesis of peptide antibiotics, in *Antibiotics,* Corcoran, J. W., Ed., Springer-Verlag, Berlin, 1981, 325.
36. **Maier, W., Erge, D., and Gröger, D.,** Zur Biosynthese von Ergotoxinalkaloiden in *Claviceps purpurea, Biochem. Physiol. Pflanzen,* 161, 559, 1971.
37. **Floss, H. G., Tcheng-Lin, M., Kobel, H., and Stadler, A.,** On the biosynthesis of peptide ergot alkaloids, *Experientia,* 30, 1369, 1974.
38. **Anderson, J. A. and Floss, H. G.,** Unpublished data.
39. **Kobel, H. and Sanglier, J. J.,** Formation of ergotoxine alkaloids by fermentation and attempts to control their biosynthesis, *FEMS Symp.,* 5, 233, 1978.
40. **Beacco, E., Bianchi, M. C., Minghetti, A., and Spalla, C.,** Directed biosynthesis of analogues of ergot peptide alkaloids with *Claviceps purpurea, Experientia,* 34, 1291, 1978.
41. **Baumert, A., Erge, D., and Gröger, D.,** Incorporation of thiazolidine-4-carboxylic acid into ergosine by *Claviceps purpurea, Plant. Med.,* 44, 122, 1982.
42. **Bianchi, M. L., Cattaneo, P. A., Crespi-Perellino, N., Guicciardi, A., Minghetti, A., and Spalla, C.,** Mechanisms of control in the quantitative biosynthesis of lysergic acid alkaloids, *FEMS Symp.,* 13, 235, 1982.
43. **Breuel, K., Braun, K., Dauth, Ch., and Gröger, D.,** Qualitative changes in peptide alkaloid biosynthesis of *Claviceps purpurea* by environmental conditions, *Plant. Med.,* 44, 121, 1982.
44. **Maier, W., Erge, D., and Gröger, D.,** Über Aktivierungsreaktionen bei *Claviceps, Biochem. Physiol. Pflanzem,* 163, 432, 1972.
45. **Maier, W. and Gröger, D.,** Einfluss von Cykloheximid auf Wachstum und Alkaloidbildung bei *Claviceps purpurea* Stamm Pepty 695, *Biochem. Physiol. Pflanzen,* 174, 1, 1979.
46. **Rao, A. S. and Singh, R.,** A speculation on the in vivo formation of bioactive oligopeptides, *Spec. Sci. Technol.,* 6, 41, 1983.
47. **Erge, D., Wenzel, A., and Gröger, D.,** Zur Physiologie der Alkaloidbildung bei Claviceps-Arten, *Biochem. Physiol. Pflanzen,* 163, 288, 1972.
48. **Stütz, P., Brunner, R., and Stadler, P. A.,** Ein neues Alkaloid aus dem Mycel eines *Claviceps purpurea* Stammes zur Biogenese von Ergocristin, *Experientia,* 29, 936, 1973.
49. **Flieger, M., Wurst, M., Stuchlík, J., and Řeháček, Z.,** Isolation and separation of new natural lactam alkaloids of ergot by high-performance liquid chromatography, *J. Chromatogr.,* 207, 139, 1981.
50. **Stuchlík, J., Cvak, L., Krajíček, A., Spáčil, J., Sedmera, P., Flieger, M., Vokoun, J., and Řeháček, Z.,** Noncyclol (lactam) ergot alkaloids, *Coll. Czechosl. Chem. Commun.,* 47, 3312, 1982.
52. **Saini, M. S., Cheng, M., and Anderson, J. A.,** Conversion of 4-γ,γ-dimethylallyltryptophan to clavicipitic acid, some properties of the enzyme, *Phytochemistry,* 15, 1497, 1976.

53. **Bajwa, R. S., Kohler, R. D., Saini, M. S., Cheng, M., and Anderson, J. A.,** Formation of clavicipitic acid in cell-free systems of *Claviceps* sp., *Phytochemistry,* 14, 735, 1975.
54. **Heinstein, P. F., Lee, S. L., and Floss, H. G.,** Isolation of dimethylallyl-pyrophosphate:tryptophan dimethylallyltransferase from ergot fungus (*Claviceps* sp.), *Biochem. Biophys. Res. Commun.,* 44, 1244, 1971.
55. **Gröger, D. and Sajdl, P.,** Enzymatic conversion of chanoclavine-I, *Pharmazie,* 27, 188, 1972.
56. **Sajdl, P. and Rehácek, Z.** Cyclization of chanoclavine-I by cell-free preparations from saprophytic *Claviceps* strains, *Fol. Microbiol.,* 20, 365, 1975.
57. **Cavender, F. L. and Anderson, J. A.,** The cell-free biosynthesis of clavine alkaloids, *Biochem. Biophys. Acta,* 208, 345, 1970.
58. **Maier, W., Schumann, B., Erge, D., and Gröger, D.,** Biosynthesis of ergot alkaloids by protoplasts of various *Claviceps purpurea* strains, *Biochem. Physiol. Pflanzen,* 175, 815, 1980.
59. **Keller, U., Zocher, R., and Kleikauf, H.,** Biosynthesis of ergotamine in protoplasts of *Claviceps purpurea, J. Gen. Microbiol.,* 118, 485, 1980.
60. **Voříšek, J. and Řeháček, Z.,** Fine structure localization of alkaloid synthesis in endoplasmic reticulum of submerged *Claviceps purpurea, Arch. Mikrobiol.,* 117, 297, 1978.
61. **Spalla, C. and Marnati, M. P.,** Aspects in the interspecific fusion of protoplasts of alkaloid producing strain of *Claviceps purpurea* and *C. paspali, FEMS,* 13, 563, 1982.
62. **Esser, K. and Tudzynski, P.,** Genetics of the ergot fungus *Claviceps paspali, Theor. Appl. Genet.,* 53, 145, 1978.
63. **Mantle, P. G. and Nisbet, L. J.,** Differentiation of *Claviceps purpurea* in axenic culture, *J. Gen. Microbiol.,* 93, 321, 1976.
64. **Řeháček, Z. and Liu, Ch.,** Unpublished data, 1982.
65. **Spalla, C. and Marnati, M. P.,** Genetics aspects of the formation of ergot alkaloids, in *Antibiotics and Other Secondary Metabolites,* Hütter, R., Leisinger, T., Nüesch, J., and Wehrli, W., Eds., Academic Press, London, 1978, 219.
66. **Weinberg, D.,** Secondary metabolism: regulation by phosphate and trace elements, *Fol. Microbiol.,* 23, 496, 1978.
67. **Taber, W. A.,** Uptake and metabolism of succinic acid by washed mycelium of *Claviceps purpurea, Mycologia,* 65, 558, 1973.
68. **Tonolo, A.,** Production of peptide alkaloids in submerged culture by a strain of *Claviceps purpurea* (Fr.) Tul., *Nature (London),* 209, 1134, 1966.
69. **Slater, E. C. and Cleland, K. W.,** The effect of tonicity of the medium on the respiratory and phosphorylative activity of heart-muscle sarcomes, *Biochem J.,* 53, 557, 1953.
70. **Atsman, A. and Davis, R. P.,** Mitochondrial respiration under conditions of varying osmolarity, *Biochim. Biophys. Acta,* 131, 221, 1967.
71. **Johnson, D. and Lardy, H.,** Substrate-selective inhibition of mitochondrial oxidations by enhanced tonicity, *Nature (London),* 181, 701, 1958.
72. **Ogunlana, E. O., Ramstad, E., and Tyler, V. E.,** Effects of some substances on ergot alkaloid production, *J. Pharm. Sci.,* 58, 143, 1969.
73. **Taber, W. A. and Vining, L. C.,** The effect of homogenizing the inoculum of *Claviceps purpurea* (Fr.) Tul., *Can. J. Microbiol.,* 3, 55, 1959.
74. **Arcamone, F., Chain, E. B., Ferretti, A., Minghetti, A., Pennella, P., Tonolo, A., and Vero, L.,** Production of a new lysergic acid derivative in submerged culture by a strain of *Claviceps paspali* Stev. Hall., *Proc. R. Soc. Ser. B,* 155, 26, 1961.
75. **Pacifici, L. R., Kelleher, W. J., and Schwarting, A. E.,** Production of lysergic acid derivatives in submerged culture. I. Fermentation studies, *Lloydia,* 25, 37, 1962.
76. **Gröger, D. and Tyler, V. E., Jr.,** Alkaloid production by *Claviceps paspali* in submerged culture, *Lloydia,* 26, 174, 1963.
77. **Řičicová, A., Flieger, M., and Řeháček, Z.,** Quantitative changes of the alkaloid complex in a submerged culture of *Claviceps paspali, Fol. Microbiol.,* 27, 433, 1982.
78. **Taber, W. A. and Vining, L. C.,** Correlation with changes in mycelial pool carbohydrate, lipid and phosphorus containing compounds, *Can. J. Microbiol.,* 9, 1, 1963.
79. **Vining, L. C. and Nair, P. M.,** Clavine alkaloid formation in submerged cultures of *Claviceps* species, *Can. J. Microbiol.,* 12, 915, 1966.
80. **Robbers, J. E., Eggert, W. W., and Floss, H. G.,** Physiological studies on ergot. Time factor influence of the inhibitory effect of phosphate and the induction effect of tryptophan on alkaloid production, *Lloydia,* 41, 120, 1978.
81. **Robbers, J. E., Robertson, L. W., Horneman, K. M., Jindra, A., and Floss, H. G.,** Physiological studies on the reduction of alkaloid synthesis by tryptophan and its inhibition by phosphate, *J. Bacteriol.,* 112, 791, 1972.

82. **Robbers, J. E., Jones, L. G., and Krupinski, V. M.,** Ergot physiology. The production of ergot alkaloids in saprophytic culture by a homokaryotic strain of ergot, *Lloydia,* 37, 108, 1972.
83. **Arcamone, F., Cassinelli, G., Ferni, G., Penco, S., Pennella, P., and Pol, C.,** Ergotamine production and metabolism of *Claviceps purpurea* strain 275 FI in stirred fermentors, *Can. J. Microbiol.,* 16, 923, 1970.
84. **Krupinski, W. M., Robbers, J. E., and Floss, H. G.,** Physiological study of ergot. Induction of alkaloid synthesis by tryptophan at enzymatic level, *J. Bacteriol.,* 125, 158, 1976.
85. **Pažoutová, S., Votruba, J., and Řeháček, Z.,** A mathematical model of growth and alkaloid production in the submerged culture of *Claviceps purpurea, Biotechnol. Bioeng.,* 23, 2837, 1981.
86. **Udvardy, E.,** Consideration of the development of an ergot alkaloid fermentation process, *Process Biochem.,* 15, 5, 1980.
87. **Řeháček, Z. and Sajdl, P.,** Changes in activity of Krebs and glyoxylate cycles during biosynthesis of agroclavine and elymoclavine, *Biotechnol. Lett.,* 1, 53, 1979.
88. **Spalla, C., Filippini, S., and Grein, D.,** A hypothesis on the regulation mechanisms governing the biosynthesis of alkaloids in *Claviceps, Fol. Microbiol.,* 23, 505, 1978.
89. **Vaidya, H. C. and Desai, J. D.,** Some metabolic studies in *Claviceps* sp. strain SD 58, *FEMS,* 13, 1982 (Poster).
90. **Pažoutová, S. and Řeháček, Z.,** The role of citrate on the oxidative metabolism of submerged cultures of *Claviceps purpurea* 129, *Arch. Microbiol.,* 129, 251, 1981.
91. **Řeháček, Z., Sajdl, P., Kozová, J., Malik, K. A., and Řičicová, A.,** Correlation of certain alternations in metabolic activity with alkaloid production by submerged *Claviceps, Appl. Microbiol.,* 22, 949, 1971.
92. **Řeháček, Z., Sajdl, P., Kozová, J., and Řičicová, A.,** Physiological activities of ergoline alkaloids in submerged cultures of *Claviceps paspali* and *C. purpurea, Fol. Microbiol.,* 17, 306, 1972.
93. **Řeháček, Z. and Liu, Ch.,** Unpublished data, 1982.
94. **Liu, Ch.,** Protein Turnover in *Saccharomyces cerevisiae* and *Claviceps purpurea,* Ph.D. thesis, Institute of Microbiology, Czechoslovakian Academy of Science, 1982.
95. **Vining, L. C. and Taber, W. A.,** Ergot alkaloids, *Econ. Microbiol.,* 3, 389, 1979.
96. **Floss, H. G. and Mothes, K.,** Über den Einfluss von Tryptophan und analogen Verbindungen auf die Biosynthese von Clavinalkaloiden in saprophytischer Kultur, *Arch. Mikrobiol.,* 48, 313, 1964.
97. **Vining, L. C.,** Effects of tryptophan on alkaloid biosynthesis in cultures of a *Claviceps* species, *Can. J. Microbiol.,* 16, 473, 1970.
98. **Schmauder, H. P. and Gröger, D.,** Zur Charakterisierung der Anthranilatsynthetase bei Claviceps, *Biochem. Physiol. Pflanzen,* 169, 471, 1976.
99. **Hornemann, U., Speedie, M. K., Kelley, K. M., Hurley, L. H., and Floss, H. G.,** Biosynthesis of indoleisopropionic acid by *Claviceps.* Biological C-methylation involving an intact methyl group, *Arch. Biochem. Biophys.,* 131, 430, 1969.
100. **Řeháček, Z. and Malik, K. A.,** Cell-pool tryptophan phases in ergot alkaloid fermentation, *Fol. Microbiol.,* 16, 359, 1971.
101. **Slater, M. L.,** Effect of reversible inhibition of DNA synthesis on the yeast cell cycle, *J. Bacteriol.,* 113, 263, 1973.
102. **Pažoutová, S., Flieger, M., Sajdl, P., and Řeháček, Z.,** The relationship between intensity of oxidative metabolism and predominance of agroclavine or elymoclavine in submerged *Claviceps purpurea* cultures, *J. Nat. Prod.,* 44, 225, 1981.
103. **Křen, V. and Řeháček, Z.,** Feedforward regulation of phosphofructokinase in a submerged culture of *Claviceps purpurea* producing clavine alkaloids, *Spec. Sci. Technol.,* 7(4), 223, 1984.
104. **Racker, E.,** History on the Pasteur effect and its pathobiology, *Mol. Cell Biochem.,* 5, 17, 1974.
105. **Atkinson, D. and Goodwin, T. W.,** *Metabolic Roles of Citrate,* Academic Press, New York, 1968, 23.
106. **El-Sayed, D. H. and Řeháček, Z.,** Variability of requirements to fermentation media in *Claviceps purpurea* strains, *Zbl. Bakt. Abt. II,* 133, 551, 1978.
107. **Řeháček, Z., Sajdl, P., Dutta, S. M., Kozová, J., and Pažoutová, S.,** Influence of Krebs cycle acids on clavine alkaloid production and changes in activity of Krebs and glyoxylate cycle enzymes during fermentation *Zbl. Bakt. Abt. II,* 135, 643, 1980.
108. **Malik, V. S.,** Genetics and biochemistry of secondary metabolism, *Adv. Appl. Microbiol.,* 28, 27, 1982.
109. **Martin, J. F.,** *Antibiotics and other Secondary Metabolism: Biosynthesis and Production,* Hüter, R., Leisinger, T., Nüesch, J., and Wehrli, W., Eds., Academic Press, New York, 1978, 19.
110. **Schmauder, H. P. and Gröger, D.,** Personal communication, 1973.
111. **Cress, W. A., Chayet, L. T., and Rilling, H. C.,** Crystallization and partial characterization of dimethylallyl pyrophosphate:L-tryptophan dimethylallyltransferase from *Claviceps* sp. SD 58, *J. Biol. Chem.,* 256, 10917, 1981.
112. **Otsuka, H., Quigley, F. R., Gröger, D., and Anderson, J. A.,** In vivo and in vitro evidence for *N*-methylation as the second pathway-specific step in ergoline biosynthesis, *Plant. Med.,* 40, 109, 1980.

113. **Yeh, L. S. and Rilling, H. C.,** Purification and properties of pig liver prenyltransferase: interconvertible forms of the enzyme, *Arch. Biochem. Biophys.,* 183, 718, 1977.
114. **Epstein, E., Ed.,** *Mineral Nutrition of Plants: Principles and Perspectives,* John Wiley & Sons, New York, 1972.
115. **Heinstein, P. and Floss, H. G.,** *Nova Acta Leoppold. Suppl.,* 7, 299, 1976.
116. **Lee, S. L.,** Ph.D. thesis, Purdue University, West Lafayette, Ind., 1974.
117. **Lee, S. L., Floss, H. G., and Heinstein, P. F.,** Purification and properties of DMAPP tryptophan dimethylallyltransferase, the first enzyme of ergot alkaloid biosynthesis in *Claviceps* sp. SD 58, *Arch. Biochem. Biophys.,* 177, 84, 1976.
118. **Erge, D., Maier, W., and Gröger, D.,** Enzymic conversion of chanoclavine-I, *Biochem. Physiol. Pflanzen,* 164, 234, 1973.
119. **Hsu, J. C. and Anderson, J. A.,** Agroclavine hydroxylase of *Claviceps purpurea, Biochim. Biophys. Acta,* 230, 518, 1971.
120. **Ogunlana, E. O., Wilson, B. J., Tyler, V. E., and Ramstad, E.,** Biosynthesis of ergot alkaloids. Enzymatic closure of ring D of the ergolene nucleus, *Chem. Commun.,* 775, 1970.
121. **Heinstein, P. F. et al.,** Unpublished data.
122. **Hsu, J. C. and Anderson, J. A.,** Personal communication, 1982.
123. **Wilson, B. J., Ramstad, E., Jansson, I., and Orrenius, S.,** Conversion of agroclavine by mammalian cytochrome P-450, *Biochim. Biophys. Acta,* 252, 348.
123a. **Kim, I. S., Kim, S. U., and Anderson, J. A.,** Microsomal agroclavine hydroxylase of *Claviceps* species, *Phytochemistry,* 20, 2311, 1981.
124. **Wiseman, A., Lim, T. K., and Wood, L. F. J.,** Regulation of the biosynthesis of cytochrome P-450 in brewer's yeast. Role of cyclic AMP, *Biochim. Biophys. Acta,* 544, 615, 1978.
125. **Ambike, S. H., Baxter, R. M., and Zahid, N. D.,** The relationship of cytochrome P-450 levels and alkaloid synthesis in *Claviceps purpurea, Phytochemistry,* 9, 1953, 1970.
126. **Ambike, S. H. and Baxter, R. M.,** Cytochrome P-450 and b_5 in *Claviceps purpurea:* interconversion of P-450 and P-420, *Phytochemistry,* 9, 1959, 1970.
127. **Řeháček, Z., Sajdl, P., and Křemen, A.,** Thermodynamic consideration in the formation of ergot alkaloids, *Biotechnol. Bioeng.,* 15, 207, 1973.
128. **Gest, H.,** *Genetics of Industrial Microorganisms,* Vaněk, Z., Hošťálek, Z., and Cudlín, J., Eds., Elsevier, Amsterdam, 1973, 131.
129. **Řeháček, Z., Kozová, J., Řičicová, A., Kašlík, J., Sajdl, P., and Švarc, S.,** Role of endogenous tryptophan during submerged fermentation of ergot alkaloids, *Fol. Microbiol.,* 16, 35, 1971.
130. **Plieninger, H., Immel, H., and Völk, A.,** Synthesis und Einbau eines ^{14}C- und ^{3}H-markierten 4-Dimethylallyl-tryptophan und eines ^{14}C-markierten 4-Dimethylallyltryptamin sowie Einbau eines ^{14}C-markierten Dimethylallylpyrophosphate, *Lieb. Ann. Chem.,* 706, 223, 1967.
131. **Řeháček, Z., Kozová, J., Sajdl, P., Kašlík, J., and Švarc, S.,** *Genetics of Industrial Microorganisms,* Vaněk, Z., Hošťálek, Z., and Cudlín, J., Eds., Elsevier, Amsterdam, 1973, 427.
132. **Cunfer, B. M. and Seckinger, A.,** *Mycologia,* 69, 1142, 1977.
133. **Turian, G., Ed.,** *Différenciation Fongigue,* Masson et Cie, Paris, 1969.
134. **Smith, J. E. and Berry, D. R., Eds.,** *Introduction to the Biochemistry of Fungal Development,* Academic Press, New York, 1974.
135. **Pasternak, C. A., Ed.,** *Biochemistry of Differentiation,* Wiley-Interscience, London, 1970.
136. **Lindegren, C. C.,** Cellular organization in relation to evolution, *J. Biol. Psychol.,* 14, 38, 1972.
137. **Lindegren, C. C.,** A speculative exploration of genetics in relation to differentiation, *Spec. Sci. Technol.,* 3, 137, 1980.
138. **Sypherd, P. S., Leathers, T., and Katayama, C.,** Molecular cloning of genes controlling the translational system of Mucor racemosus, ASM Conf. Gene Manipulation in the Exploration and Study of Fungi, Abstr. p. S-11, South Bend, Ind., 1983.
139. **Smith, J. E. and Galbraith, J. C.,** *Advances in Microbial Physiology,* Vol. 5, Rose, A. H. and Wilkinson, J. F., Eds., Academic Press, London, 1971.
140. **Turian, G.,** Reducing power of hyphal tips and vegetative apical dominance in fungi, *Experientia,* 32, 989, 1976.
141. **Zalokar, M.,** Enzyme activity and cell differentiation in *Neurospora, Am. J. Bot.,* 46, 555, 1959.
142. **Turian, G.,** Differentiation in *Allomyces* and *Neurospora, Trans. Br. Mycol. Soc.,* 64, 367, 1975.
143. **Mantle, P. G. and Tonolo, A.,** Relationship between the morphology of *Claviceps purpurea* and the production of alkaloids, *Trans. Br. Mycol. Soc.,* 51, 499, 1968.
144. **Voříšek, J., Ludvík, J., and Řeháček, Z.,** The morphogenesis and ultrastructure of *Claviceps purpurea* during alkaloid formation, *J. Bacteriol.,* 120, 1401, 1974.
145. **Voříšek, J. and Řeháček, Z.,** Origin of sclerotia-like cells in submerged *Claviceps purpurea* producing clavine alkaloids, *Arch. Microbiol.,* 107, 321, 1976.

146. **Komersová, I., Ludvík, J., and Majer, J.,** The ultrastructure of spores of *Claviceps purpurea* (Fr.) Tul., *Fol. Microbiol.*, 14, 141, 1969.
147. **Schmauder, H. P., Gröger, D., Lösecke, W., Rudolph, A., and Grimmecke, H. D.,** Über die chemische Zusammensetzung der Zellwant von *Claviceps*-Arten, *Biochem. Physiol. Pflanzen,* 173, 295, 1978.
148. **Bauer, H. and Sigarlakie, E.,** Localization of alkaline phosphatase in *Saccharomyces cerevisiae* by means of ultrathin frozen sections, *J. Ultrastruct. Res.*, 50, 208, 1975.
149. **Voříšek, J. and Pokorný, V.,** Cytochemical detection of polysaccharides on the surface of the cell membrane complex in fungi, *Arch. Microbiol.*, 102, 293, 1975.
150. **Corbett, K., Dickerson, A. G., and Mantle, P. G.,** Metabolic studies on Claviceps purpurea during parasitic development on rye, *J. Gen. Microbiol.*, 84, 39, 1974.
151. **Pažoutová, S., Pokorný, V., and Řeháček, Z.,** The relationship between oxidation and production of alkaloids in saprophytic strains of *Claviceps purpurea, Can. J. Microbiol.*, 23, 1182, 1977.
152. **Řeháček, Z., Kozová, J., Sajdl, P., and Voříšek, J.,** The physiology of conidia formation in submerged culture of *Claviceps purpurea* (Fr.) Tul. producing alkaloids, *Can. J. Microbiol.*, 20, 1323, 1974.
153. **Pažoutová, S., Řeháček, A., and Voříšek, J.,** Induction of sclerotia-like mycelium in saprophytic *Claviceps purpurea* producing clavine alkaloids, *Can. J. Microbiol.*, 26, 363, 1980.
154. **Pažoutová, S. and Řeháček, Z.,** Unpublished data.
155. **Dickerson, A. G., Mantle, P. G., and Sczyrbak, C. A.,** Autolysis of extracellular glucans produced in vitro by a strain of *Claviceps fusiformis, J. Gen. Microbiol.*, 60, 403, 1970.
156. **Voříšek, J. and Sigler, K.,** Electron-cytochemical study of lipid deposit condensation and degradation in ascomycetes-self-assembly of unit membranes, *Cell. Mol. Biol.*, 28, 221, 1982.
157. **Pöhm, M., Kolassa, H., and Jentsch, K.,** Zur Kenntniss der proteolytischen Aktivität des Mutterkorns, *Sci. Pharmaceut.*, 44, 32, 1976.
158. **Kaplan, H., Hornemann, V., Kelley, K. M., and Floss, H. G.,** Tryptophan metabolism, protein and alkaloid synthesis in saprophytic cultures of the ergot fungus (*Claviceps* sp.), *Lloydia,* 32, 489, 1969.
159. **Roth, K. and Fritsche, W.,** Beziehung zwischen Proteinsynthese und Alkaloidbildung beim Mutterkornpilz in saprophytischer Kultur, *Arch. Mikrobiol.*, 58, 77, 1967.
160. **Řeháček, Z., Desai, S. D., Sajdl, P., and Pažoutová, S.,** The cellular role of nitrogen in the biosynthesis of alkaloids by submerged culture of *Claviceps purpurea* (Fr.) Tul., *Can. J. Microbiol.*, 23, 596, 1977.
161. **Tonolo, A.,** Alcune considerazioni sulla produzione di alcaloidi in coltore artificiali con specie diverse di *Claviceps* Tul., *Ann. Inst. Super. Sanita,* 3, 613, 1967.
162. **Puc, A. and Sočič, H.,** Carbohydrate nutrition of *Claviceps purpurea* for alkaloid production, related to the osmolarity of media, *Eur. J. Appl. Microbiol.*, 4, 283, 1977.
163. **Neumann, D., Lösecke, W., Maier, W., and Gröger, D.,** Localization of alkaloids in sclerotia and suspension culture of *Claviceps purpurea* (Fr.) Tul, *Biochem. Physiol. Pflanzen,* 174, 504, 1979.
164. **Streiblová, E., Rýc, M., and Kybal, J.,** Fine structure of imbided sclerotial cells of *Claviceps purpurea* (Fr.) Tul revealed by freeze-etching, *Z. Allg. Mikrobiol.*, 18, 124, 1978.
165. **Rest, J. A. and Vaugham, J. G.,** The development of protein and oil bodies in the seed of *Sinapis alba, Planta,* 105, 245, 1972.
166. **Yatsu, L. Y., Jacks, T. J., and Heusarling, T. P.,** Isolation of spherosomes (oleosomes) from onion, cabbage and cotton-seed tissue, *Plant. Physiol.*, 48, 675, 1971.
167. **Hislop, E. C., Barnaby, V. M., Shellis, C., and Laborda, F.,** Localization of α-arabinofuranosidase and acid phosphatase in mycelium of *Sclerotinia fructigena, J. Gen. Microbiol.*, 81, 79, 1974.
168. **Voříšek, J., Sajdl, P., and Řeháček, Z.,** Localization of alkaloid synthesis in submerged *Claviceps purpurea* — a cytochemistry study, 3rd Symp. Soc. Countries on Biotechnology, Bratislava, 1983 (Poster).
169. **Řeháček, Z. and Kozová, J.,** Production of alkaloids and differentiation in submerged culture of *Claviceps purpurea* (Fr.) Tul, *Fol. Microbiol.*, 20, 112, 1975.
170. **Řeháček, Z.,** Physiology of Formation of some Antibiotics and Ergot Alkaloids, D.Sc. thesis, Institute of Microbiology, Czechoslovakian Academy of Science, Prague, 1972.
171. **Řeháček, Z.,** Ergot alkaloids and some problems of the physiology of their formation, *Zbl. Bakt. II,* 129, 20, 1974.
172. **Řeháček, Z.,** Ergot alkaloids and their biosynthesis, *Adv. Biochem. Eng.*, 14, 33, 1980.
173. **Schwarz, G. and Eich, E.,** Influence of ergot alkaloids on growth of *Streptomyces purpurascens* and production of its secondary metabolites, *Plant. Med.*, 47, 212, 1983.
174. **Arnison, P. G.,** A redox model of the mechanism of action of indole acetic acid (auxin) and other plant growth regulators, *Spec. Sci. Technol.*, 3, 5, 1980.
175. **Champoux, J. J.,** Proteins that affect DNA conformation, *Ann. Rev. Biochem.*, 47, 449, 1978.
176. **Birch, A. J., McLoughlin, B. J., and Smith, H.,** The biosynthesis of the ergot alkaloids, *Tetrahedron Lett.*, 7, 1, 1960.
177. **Taylor, E. M. and Ramstad, E.,** Biosynthesis of lysergic acid in ergot, *Nature (London),* 188, 494, 1960.
178. **Tyler, V. E.,** Biosynthesis of the ergot alkaloids, *J. Pharm. Sci.*, 50, 629, 1961.

179. **Cheng Li-Jung, Robbers, J. E., and Floss, H. G.,** End-product regulation of ergot alkaloid formation in intact cells and protoplasts of *Claviceps* species, strain SD 58, *J. Nat. Prod.*, 43, 329, 1980.
180. **Řeháček, Z. and Basappa, S. C.,** Influence of Tween 80 on the production of ergot alkaloids by submerged culture of *Claviceps paspali*, *Fol. Microbiol.*, 16, 110, 1971.
181. **Demain, A. L.,** Regulatory mechanism and the industrial production of microbial metabolites, *Lloydia*, 31, 395, 1968.
182. **Hütter, R., Niederberger, P., and Aebi, M.,** Regulation of amino acid biosynthesis in eucaryotic microorganisms, *Abstr. 4th Int. Symp. GIM*, Tokyo, 1982, 7.
183. **Fink, G. R., Lucchini, G., and Hinnebusch, A. G.,** Repeated DNA sequences and regulatory genes that control biosynthetic genes in yeast, ASM Conf. Gene Manipulations in the Exploitation and Study of Fungi, Abstr. S-7, South Bend, Ind., 1983.
184. **Esser, K., Kück, V., Stahl, U., Tudzynski, P.,** A new concept in genetic engineering, ASM Conf. Gene Manipulation in the Exploitation and Study of Fungi, Abstr. S-5, South Bend, Ind., 1983.
185. **Tyler, V. E., Erge, D., and Gröger, D.,** Biological conversions of agroclavine and elymoclavine, *Plant. Med.*, 13, 315, 1965.
186. **Voigt, R., Bornschein, M.,** Zur Stellung des Chanoclavins im Alkaloid-stoffwechsel von *Claviceps purpurea* Tul., *Pharmazie*, 19, 342, 1964.
187. **Eich, E.,** Zur Steigerung der Production von Lysergsäureamiden in Submerskulturen von *C. purpurea*, *Plant. Med.*, 23, 330, 1973.
188. **Vining, L. C.,** Conversion of alkaloids and nitrogenous xenobiotics, in *Economic Microbiology*, Vol. 5, Rose, A. H., Ed., Academic Press, London, 1980, 523.
189. **Agurell, S. and Ramsted, E.,** Biogenic interrelationships of ergot alkaloids, *Arch. Biochem. Biophys.*, 98, 457, 1963.
190. **Lin, W. N. C. and Ramstad, E.,** Enzymology of ergot alkaloid biosynthesis. III. 10-Hydroxyelymoclavine, an intermediate in the peroxidase conversion of elymoclavine to penniclavine and isopenniclavine, *Lloydia*, 30, 202, 1967.
191. **Taylor, E. H. and Shough, H. R.,** Enzymology of ergot alkaloid biosynthesis. II. The oxidation of agroclavine by horseradish peroxidase, *Lloydia*, 30, 197, 1967.
192. **Yamatodani, S., Kozu, Y., Yamada, S., and Abe, M.,** *Annu. Rep. Takeda Res. Labs*, 21, 88, 1962.
193. **Abe, M., Abh. Dtsch. Akad. Wiss.,** Berlin, *Kl. Chem. Geol. Biol.*, 3, 393, 1966.
194. **Brack, A., Brunner, R., and Kobel, H.,** Mikrobiologische Hydroxylierungen an Mutterkornalkaloiden vom Claviceps-Typus mit dem mexikanischen Rauschpilz Psylocybe semperviva, Heim et Cailleux, *Helv. Chim. Acta*, 45, 276, 1962.
195. **Abe, M.,** *Abh. Dtsch. Akad. Wiss.*, Berlin, 4, 309, 1963.
196. **Floss, H. G.,** Biosynthesis of ergot alkaloids and related compounds, *Tetrahedron Lett.*, 32, 873, 1976.
197. **Ramstad, E., Chan-Lin, W. N., Shough, H. R., Goldner, K. J., Parikh, R. P., and Taylor, E. H.,** Norsetoclavine, a new clavine-type alkaloid from Pennisetum ergot, *Lloydia*, 30, 441, 1967.
198. **Cassady, J. M., Abou-Chaar, C. I., and Floss, H. G.,** Ergot alkaloids. Isolation of *N*-demethylchanoclavine-II from *Claviceps* strain SD 58 and the role of demethylchanoclavines in ergoline biosynthesis, *Lloydia*, 36, 390, 1973.
199. **Schmauder, H. P.,** Saphrophytic production of ergot alkaloids, in Cultivation and Utilization of Medicinal Plants, Atal, C. K. and Kapur, B. M., Eds., Reg. Res. Lab., Council Sci. Ind. Res., Jammu-Tawi, 1982, 188.
200. **Banks, G. Z., Mantle, P. G., and Sczyrbak, C. A.,** Large-scale production of clavine alkaloids by *Claviceps fusiformis*, *J. Gen. Microbiol.*, 82, 345, 1974.
201. **Thom, R., Ed.,** *Structural Stability and Morphogenesis*, W. A. Benjamin, Reading, Mass., 1975.
202. **Demain, A. L.,** Microbial production of primary metabolites, *Naturwiss.*, 67, 582, 1980.
203. **Kobel, H., Brunner, R., and Brack, A.,** Vergleich der Alkaloidbildung von *Claviceps purpurea* in parasitischer und saprophytischer Kultur, *Experientia*, 18, 140, 1962.
204. **Kobel, H. and Sanglier, J. J.,** Qualitative changes in the alkaloid spectrum of *Claviceps purpurea* after mutation, in *Proc. 1st Int. Symp. Genetics of Ind. Microorganisms*, Vol. 2, Vaněk, Z., Hošťálek, Z., and Cudlín, J., Eds., Elsevier, Amsterdam, 1973, 421.
205. **Srikrai, S. and Robbers, J. E.,** Methods for mutation and selection of the ergot fungus, *Appl. Environ. Microbiol.*, 45, 1165, 1983.
206. **Dykhuizen, D. E. and Hartl, D. L.,** Selection in chemostat, *Microbiol. Rev.*, 47, 150, 1983.
207. **Mantle, P. G.,** in *Filamentous Fungi*, Vol. 1, *Industrial Mycology*, Smith, J. E. and Berry, D. R., Eds., Edward Arnold, London, 1974, 281.
208. **Tudzynski, P., Düvell, A., and Esser, K.,** Extrachromosomal genetics of, *Claviceps purpurea*, *Curr. Genet.*, 7, 145, 1983.
209. **Tudzynski, P. and Esser, K.,** *Curr. Genet.*, 6, 153, 1982.
210. **Drews, S. W. and Demain, A. L.,** Effect of primary metabolites on secondary metabolism, *Ann. Rev. Microbiol.*, 31, 343, 1977.

211. **Crick, F. H.,** Codon-anticonon pairing, the wobble hypothesis, *J. Mol. Biol.,* 19, 548, 1966.
212. **Watson, J. D., Ed.,** *Molecular Biology of the Gene,* 3rd ed., W. A. Benjamin, Menlo Park, Calif., 1976.
213. **Kobel, H.,** Bildung von Ergotoxinalkaloiden in Mischkultur. Conf. Med. Plants, Mariánské Lázně, Abstr. p. 72, 1975.
214. **Hanson, R. S.,** Role of small molecules in regulation of gene expression and sporogenesis in bacilli, in *Spores,* Vol. 6, Gerhardt, P., Costilov, R. N., and Sadoff, H. L., Eds., ASM, Washington, D.C., 1975, 318.
215. **Taber, W. A.,** Fermentative production of hallucinogenic indole compounds, *Lloydia,* 30, 39, 1967.

INDEX

A

D

E

F

G

H

I

K

L

M

N

O

P

R

S

U

V

Y

Z